NEW YORK POWER

by Joseph J Cunningham

Copyright © 2013 by Joseph J Cunningham
All rights reserved.
ISBN: 1484826515
ISBN 13: 9781484826515
Library of Congress Control Number: 2013908868
CreateSpace Independent Publishing Platform
North Charleston, South Carolina

No part of this work may be reproduced, stored in any retrieval system, or transmitted in any form or by any means, electronic, mechanical, photocopying, microfilming, recording, or otherwise, without express written permission from the author.

Cover photo: Power Station concept drawing Thomas E Murray
(Scanned & Photo-Shopped by R.W. Lobenstein)

DEDICATION

To grandpa Al who lived the pioneer electric era; and to my mom Isabel, who kept his legends alive for me. Also to all those who have worked or contributed to power operations in any way, anywhere; from the first Brush and Edison employees to those at Con Ed and all the manufacturers and other companies whose labor keeps the lights on and the power flowing.

Contents

Acknowledgments....................................vii
Foreword..xiii
Introduction......................................xvii
I: Pioneers Introduce Private Power................1
II: The Central Stations............................9
III: Alternating Current Generation and
      Transmission 37
IV: Growth Through Economy of Scale................61
V: Direct Current Customer Service Peak...........91
VI: Alternating Current Incursion.................107
VII: Alternating Current Triumph = Direct Current
      Demise..145
VIII: Consolidation.................................159
IX: Postwar Prosperity............................169
X: Recent Change and Developments................185
Bibliography......................................195
Index...209

Photo Credits:

1 through # 3: Photo Album *N Y Edison Co. Properties* 1905 Scanned and Photo-Shopped by R.W. Lobenstein

4 through # 7: *Murray, Thomas E. Electric Power Plants; A Description of A Number of Power Stations. New York, 1910* – Scanned and Photo-Shopped by R.W. Lobenstein

#8 and #9: Electrical World May 1911 Scanned by S. Hochheiser Photo-Shopped by L.O. DeHart

#10: R.W. Lobenstein photo

Rear Cover: Author at Disconnection of last Direct Current Customer November 14, 2007 -R.W. Lobenstein photo

ACKNOWLEDGEMENTS

A multitude of people contributed to this book, unfortunately there is no way to list them all so I will try to cover the highlights. From my youngest days there were the Con Ed workers outside a power station, an excavation, or structure who took time from their lunch to answer the questions of a curious boy. An inspector by the name of McFadden, called to address my family's complaint of power theft by a neighbor, took time to explain to me how the distribution system worked, at least at a level understandable by a fifth grader.

This book had its inception in the mid 1990s when I decided to do some research on the old direct current networks. Thirty years earlier I had devoured every book on electrical history that I could find in fifth and sixth grade, but the standard description of the "War of the Currents" just didn't ring true even then. I knew that alternating current did not abolish direct current instantly or immediately as the tales would have one believe. We knew people and places where direct current was still in operation in the 1950s and early 1960s. One of my earliest memories on the topic was an August 17, 1959 power failure when older apartments on the Central Park networks remained lit with direct current while the stores and businesses below were dark, having been modernized with alternating current.

An inquiry to Con Ed led me to Steven Jaffe, the company librarian who found for me an old folio of articles from the *Edison Weekly* that described the direct current system of the 1920s. While discussing a totally unrelated issue with Bill Wall of the Transit Authority, he mentioned that he had some old books purchased years earlier for 25 cents when he worked for Con Ed. Those happened to be the legendary books of Thomas E. Murray, the pivotal figure in New York electrification but one then unknown to myself and most others.

About the same time I met Robert Lobenstein (Loby) now retired General Superintendent of Power Operations and also the unofficial historian of the Transit power department. We began discussing and sharing data – in time he found a master list of all the old Edison direct current substations and we began a series of "Substation Safaris" with his Jeep exploring the sites of former Edison facilities. Some were empty lots, some converted to other purposes. The most interesting site was a recording studio in which the technical staff had inquired of Loby about the original purpose of the building. They gave us a cellar-to-penthouse tour— provided that we were clear by noon because the

ACKNOWLEDGEMENTS

legendary musician Roberta Flack was scheduled for a session later that day. At another site, a gentleman approached us with his card – seems the property was for sale! A good deal but what would we do with it – not to mention the taxes!

If there were a formal research contributor on this work it would be Loby, for he put me in touch with numerous people at Con Ed with whom he had worked over the years. Bob Mills, Ed Pope, Frank Baresi, John O'Malley, John McGregor, and Mike Kessler, all of whom shared stories about remnants of the old systems. Louis Rana, then VP of Systems and Transmission (later CEO) was referred to me when he was preparing a presentation on the retirement of the old 25 Hz system. It was he who revealed to me during that phone conversation that the old direct current system was still active in 2000 - much to my surprise.

After I conceived the idea of a history of power, numerous people encouraged me to do it, especially Bob Walker, retired Chief Engineer of Metro-North Railroad with twenty-eight years in the railroad power department, Peter Derrick and Gary "Doc" Hermalyn of the Bronx County Historical Society, and others with an interest in power history. Peter Dietrich of the Con Ed Learning Center provided access to ancient typewritten lists of the changeover from direct to alternating current in Manhattan in the 1930s. Michael Wares, librarian at Fordham, pointed me in the direction of interesting and very useful volumes.

Intensive research at the Science, Industry and Business Library of the New York Public Library brought out much of the background including the importance of Thomas E. Murray and the role of the United Electric Light & Power Co. the alternating current pioneer whose role has been largely forgotten. The length and complexity of the power story in New York City soon made it obvious that it is just too much to fit in one volume and that there are just too many unknowns in the history of power in Queens, Staten Island and Westchester to permit of a really comprehensive work. So I decided that the only practical approach was to focus on the history of the Manhattan system, for it has been recorded meticulously. Of that, the midtown area—the most concentrated electrical load center in the world—has the most dynamic history.

Jackie Gold, now retired from the library, gave me an opportunity to do a series of presentations which brought me into contact with the IEEE and Con Ed staff from which I learned more. Thanks to Bob McGee of the Con Ed media relations department, I was able to be at a contemporary event of midtown electrical history, the final cable cut that ended direct current distribution in 2007.

Despite the interest on the part of electrical professionals, it was obvious that ordinary sales channels would never justify the expense of publication of the book, so I satisfied myself with some articles in the *IEEE Power & Energy Magazine* in the belief that my

2001-2003 manuscript was destined for existence only as ones and zeros on computer disks.

I did extract parts of it for presentations to various groups. Along the way I got to know— either in person or by e-mail—a host of people who were interested in power history, a list that includes but is by no means limited to: Dan Taft, Ken Vought, Shu-Ping Chang, Bob Pellegrino, Carl Sulzberger, Mel Olken, Alex Magoun, Sheldon Hochheiser, Michael Geselowitz, Neil Weisenfeld, and Arnold Wong. Tom Blalock, a fellow contributor to *IEEE Power & Energy Magazine* , provided some very useful leads to additional data. I am sure I have forgotten some people – hope they won't resent it too much!

Then one day, Carl Sulzberger, a great guy and the history editor of the *IEEE Power & Energy Magazine*, who had given me a thumbs up to every idea I floated, mentioned the new print-on-demand book program being launched by the IEEE History Center. Perhaps it wasn't a dead issue after all! After discussion with some other contacts, especially Robert Colburn whom I knew from the IEEE Milestones in Electrical Engineering and Computing Program, it came about.

Finally, while I have had much help and encouragement from people at Con Ed, this should, in no way, be interpreted as anything official or endorsed. It is my research, writing, and yes—in the last chapter— some of my opinion. While I thank all of the Con Ed people who offered assistance; any errors, omissions, misinterpretations, or general foul-ups are mine for which I take full responsibility.

Though by no means a complete electrical or corporate history, it is my sincere hope that it will honor the past inventors, builders and workers; salute the magnificent work of those who keep it running today, and inspire all those who will have to meet the challenges and endure the struggles of the future.

Joseph J. Cunningham
December 2012

FOREWORD

In writing *New York Power*, author Joseph J. Cunningham has consulted a wide variety of historical sources to bring us the story of the development of today's New York City electric utility system. To fully appreciate the importance of the story told in this book, the reader must understand that, from the earliest days of electrification to today, the electrical load density (electric power requirements per unit of area) in midtown Manhattan, New York City, has been greater than anywhere else in the world. As a result, a great deal of vision, innovation, engineering, and investment have been devoted to developing and expanding the electric utility system in Manhattan. In a number of ways, the massive and sustained effort to provide adequate, reliable, and economical electric utility service to Manhattan and, over time, to the other four boroughs that are also part of New York City was a prototype for electric utility systems elsewhere.

Further, the early and intense competition in North America between direct current (dc) electric power and alternating current (ac) electric power for supplying the growing residential, commercial, and industrial development was largely played out in New York City, particularly in Manhattan. Thomas Edison was the first to enter the arena in 1882 with his well known pioneering Pearl Street dc central generating station and its dc electric power distribution system. By the late 1880s, ac electric power systems, championed by the industrialist George Westinghouse, began to appear.

Most written accounts of this dc versus ac electric power rivalry focus on the advances made by both dc and ac systems during the decades of the 1880s and 1890s. These accounts usually conclude with a short paragraph noting that, due to its inherent advantages, ac power systems prevailed and dc power systems experienced inevitable decline. However, what followed these two decades was a long, arduous, and costly effort lasting well into the twentieth century before ac electrical systems gained general acceptance and widespread application, thereby supplanting dc for the provision of electric utility service.

In addition to describing the many advances in electric power technology from the late 1880s to today, Cunningham also discusses the talented pioneering engineers and businessmen who created and made use of electrotechnology to meet the electric power needs of an ever-growing metropolis.

FOREWORD

Joe Cunningham has enjoyed a life-long interest in electric power systems. When he was in high school, his science project on "The Theory and Operation of Alternating Current" was awarded a first place gold medal, and this success led to a scholarship for the study of physics in college. Cunningham has researched and authored numerous brochures, articles, and books on topics such as industrial electrification, electric utility power systems, and electric rail transportation. He has also lectured and taught widely on the history of electric power systems and has consulted on numerous electrotechnology projects and television productions.

New York Power is a well researched and informative resource that should be of great interest to electric utility engineers. students, and anyone else wishing to learn more about the rich history of electric power and the visionary engineers and entrepreneurs who made it all happen.

Carl L. Sulzberger
Associate Editor, History
IEEE Power & Energy Magazine

INTRODUCTION

New York City has long represented one of the most concentrated urban developments in the world. That density has placed unique constraints on every aspect of life. Electric light and power appeared during the 1880s, but much development was required to supply urban service at a cost that would make possible large-scale consumption. Innovation was needed most in midtown Manhattan, where the sheer density of electrical load overwhelmed the early systems and which continues to be the greatest concentration of electrical load in the world. The first public service was initiated in 1880 with the illumination of Broadway, Madison Park and some businesses by arc lights of the Brush Electric Company. Two years later, Thomas Edison introduced incandescent light service to the offices and businesses of the financial district from his station on Pearl Street. While that installation entered the record books, his long term objective was the midtown area. Edison considered the establishment of electric service in the area of the West Twenties and Thirties vital to the future of his company.

He succeeded, but the limited transmission range of the direct current system in use at that time presented an obstacle to large scale electric service. It was obvious that the load of the midtown area required electric capacity on a scale that surpassed any planned elsewhere. Furthermore, the electric demand of the area was constant as the city became a twenty-four hour metropolis. The daytime load of industrial and commercial activity was supplanted in the evening by that of theaters, hotels and restaurants. As those loads and that of residential customers tapered off, the demand of bakeries, dairies, cleaning services, and other nighttime businesses reached a peak.

Pioneer alternating current systems promised to make large scale installations feasible, but they required substantial research and development. Some experts favored the use of alternating current exclusively. Others preferred the retention of direct current distribution with the use of alternating current restricted to generation and transmission. The latter approach predominated as a result of technical constraints and the substantial investment in direct current motors that were in place though a small but determined competitor introduced an alternating current system to the area. In 1926, the innovators succeeded and the decision to retire direct-current service followed two years later. The size and complexity of the "changeover" process was such that it took decades, and remnants of the old direct current distribution system were not retired until late 2007.

Evolution has been constant as new concepts and techniques have been developed. The local system has been interconnected with those of other companies, states, and Canada to produce a regional network of power supply. All that effort has been directed toward one objective: the provision of reliable and efficient service to an area that has no equal in the degree of load density. New networks are installed, systems rearranged, and capacity increased even as the topography of the city evolves. This book will focus on the midtown area from Times Square to Madison Square that was the site of the city's first electric street illumination, and also that of the first power stations planned to provide a full range of service. It was also the area in which alternating current first challenged and then ultimately vanquished the original direct current system.

I: PIONEERS INTRODUCE PRIVATE POWER

The industrial revolution that swept the United States after the Civil War produced overcrowding in Manhattan that threatened the economic stability of the city. Confined to a long narrow island, the city could expand only to the north. Elevated railway lines enabled such expansion for residential development, but business favored the midtown and downtown locations. In those areas, the only practical expansion was upward. Steel "skeleton" frame construction promised to expand the height of structures but that option was constrained by the limited ability of citizens to tolerate multiple flights of stairs. Hydraulic elevators were developed but proved cumbersome and expensive. As a result, the highest points in the city were church steeples and the Brooklyn Bridge.

At that time, the Madison Park area was one of residences and fine hotels about to be overtaken by commercial development. By 1880, it was the northern end of the

"Ladies' Mile" of specialty shops and department stores that advanced along Sixth Avenue and Broadway from the vicinity of Eighth Street. North of Madison Park, the residences gave way to an emerging entertainment district that later became known as "Tin Pan Alley." The two year-old elevated railway on Sixth Avenue was considered a double-edged sword; it brought commerce but also disrupted the genteel life. The area of the West Thirties known as the "Tenderloin" was declared by civic reformers to be the most morally decadent spot on earth. A number of schemes were proposed for what would today be called "redevelopment" but large-scale change required increased industry and commerce.

To do so, it was necessary to develop improved power sources for industry and transportation. Coal fired steam engines were used extensively, nevertheless progressive researchers sought to develop electric power. Thomas Davenport of Vermont had patented a basic electric motor in 1835, and the Baltimore & Ohio Railroad tested an electric locomotive in 1851. Both efforts used the chemical batteries of the day. The limited capacity of those batteries proved an obstacle to the further development of electric motors. Most research effort focused on the development of electric light, and those pioneers built the first systems for mechanical generation of electric power.

The initial efforts to produce an electric light used an electric arc (spark) sustained between carbon electrodes. Explored initially for lighthouses and naval

I: PIONEERS INTRODUCE PRIVATE POWER

searchlights, the arc light was developed subsequently for the illumination of streets and large interior spaces. The leading arc light pioneer in the United States was Charles Brush of Cleveland, Ohio. After an initial demonstration in his home, Brush arc lamps were installed in San Francisco in 1879 and then at the courthouse at Wabash, Indiana, the following year. The 1879 installation by the California Electric Light Company in the Pacific Building in San Francisco supplied lights in other buildings and may have been the first instance of public electric service. If the power line crossed a public thoroughfare, then that installation would qualify as the first public electric utility service under one standard definition.

Little has been recorded, but it is known that a substantial arc light installation was located in the College of St. Ignatius, now the University of San Francisco. The work of Jesuit Father Joseph Neri, it was part of the Cabinet (Department) of Physical Sciences, but was expanded temporarily along Market Street in 1876 as part of the celebration of the nation's Centennial on July 4th of that year. More than a decade ahead of his contemporaries, Fr. Neri encouraged development of electric power with the declaration that it would one day revolutionize industry and rail transportation.

Professor Elihu Thomson of Philadelphia was another leader in the research effort, having developed an arc lamp independently of Brush. An installation in a local bakery saved the lives of delivery wagon horses during a disastrous fire for, in the words of the fire chief, "it remained lit no matter how much water poured on it." Thomson and his colleague Edwin Houston then formed the Thomson-Houston Company to market the system. A host of other inventors and electric companies marketed arc light systems by the early 1880s.

All of those installations required substantial investment and innovation because no power system existed at that time. Mechanical power generation was mandatory as chemical batteries could not produce current at sufficient voltage (pressure) to operate an arc light system. Early efforts to produce mechanical generators encountered an obstacle in that the rotation of the coils produced a current that oscillated in both strength and direction. Incompatible with the steady unidirectional *direct current* produced by chemical batteries, the *"alternating"* current produced mechanically did not obey the simple rules that had been derived by the physicists of the period. An automatic switch or *commutator* was developed that reversed the connections as needed to produce a steady direct current. That invention made possible a practical direct current generator.

One of the first practical machines was invented by James J. Wood, a young factory foreman in Brooklyn. Then known as *dynamos*, such generators were exhibited at the Philadelphia Centennial Fair of 1876, but the Bell telephone and Corliss stationary steam engine attracted greater public interest. Prof. Thomson was engaged by the Franklin Institute to survey and test a variety of

1: PIONEERS INTRODUCE PRIVATE POWER

generators prior to the purchase of one by the museum. Thomson produced an extensive catalog of the available designs which proved to be of great value to all of the pioneers. It was also necessary to use a steam engine that would perform well when connected to an electric generator. Speed control was vital as variation produced instability in the voltage and thus in the intensity of the light that was produced.

Those pioneer arc light systems of Brush and Thomson were the first of a new technology that transformed the interiors of institutions and large industrial and commercial establishments. Power was supplied by a *lighting plant* that consisted of an electric generator driven by a steam engine. Most were installed in basements; large institutions often used separate structures to isolate the noise and smoke. Arc lamps were limited to large spaces because the glow was much too brilliant for direct viewing in a confined area. A number of researchers sought to reproduce the soft glow of gas mantles in an electric light suitable for homes and offices.

Soft Light

Most of the efforts to develop an electric light suited to interior spaces entailed the use of a material heated to incandescence by internal resistance to the passage of the current. Thomas Edison, an inventor already famed for his phonograph, stock ticker, and telegraphy inventions, received the patent that was later recognized as primary. His first successful lamp was produced in October, 1879. Edison then developed a complete lighting system and announced a demonstration at his laboratory in Menlo Park, N.J. Held on New Year's Eve; the public response was such that the Pennsylvania Railroad had to operate special trains to carry the throngs of visitors. In addition to lights in the buildings, the grounds were illuminated by lamps suspended from wires. Each lamp was rated at sixteen candlepower equivalent (c.p.) and consumed about 56 watts. The system operated at 110 volts, a pressure that simplified the design of generators, wires, and lamps.

With the Thomson report as a base, Edison directed his staff to develop a series of generators rated in size from seventy-five to two-hundred and fifty lamps. Edison lighting plants were soon in service not only in homes and offices, but also on ships and railroad cars. Competitors sold similar lamps and lighting systems, some of which infringed upon the Edison patent. Edison preferred to avoid litigation by allowing the marketplace to settle the issue. That approach failed, and extensive litigation later proved the Edison patent to be primary. Some of the contenders had valid claims. Joseph Swan of England invented the lamp independently, and Brush had obtained rights to the Swan patents in an effort to insure the continued viability of his company. In England, the Edison and Swan interests were combined in an enterprise named Ediswan.

I: PIONEERS INTRODUCE PRIVATE POWER

Edison promoted his light with vigor; in 1884 the company sponsored an electric light parade down Fifth Avenue. Hidden generators on "floats" supplied a number of lamps on the hats and uniforms of the marchers, even the baton of the leader was adorned with a lamp!

The convenience and appearance of electric lighting was obvious to all, but was affordable only by the wealthy. Still, electric lighting was popular and the incandescent light found application in the better hotels and successful business establishments. The sound of throbbing steam engines represented progress to urban dwellers familiar with steam engines in factories. The elimination of noxious fumes from gas seemed a wonder; the nuisance of the plants a small price to pay for progress. Most were located in basements; the repair of such a plant was said to be the first work undertaken by the brilliant Serbian electrical scientist Nikola Tesla upon his arrival in New York. Observing the emanation of blue flashes and language of similar hue from a basement, Tesla found a man frustrated by a cantankerous generator. Familiar with the machine from his work in Europe, Tesla repaired the machine in short order and made a business contact that proved significant to his later endeavors.

As electric light became established, novelty items appeared. Festive electric lighting was introduced on the evening of Friday, December 22, 1882 when Edison associate Edward H. Johnson opened his home to visitors to display the world's first electrically lighted Christmas

Tree. The leading trade journal, Electrical World, described in January of 1885 a Christmas tree in the home of Johnson. Festooned with small lamps of different colors that flashed sequentially as the tree was rotated by a motor in the base, it was a sensation of the day.

Brush, Edison, Thomson, and others anticipated a huge market for electric light service sold as a utility in the same manner as that by which gas, telegraph, and telephone services were provided. That required the development of a central station to generate power and a distribution system to supply individual customers. In that way the expense of construction and operation might be divided among thousands of users. Such public service required not only major financial investment but also substantial technical development. As the arc light companies enjoyed a substantial lead, those systems were the first to provide public service.

II: THE CENTRAL STATIONS

Brush anticipated the potential market in New York City as vital to the large scale application of his lighting system. In view of the advance of the Ladies' Mile and the northward movement of business, Brush selected Broadway in midtown for the first installation. The first electric power station known to have provided public service was constructed by the Brush Electric Company just west of Sixth Avenue at 133 West 25th Street. Power was carried by overhead lines at a pressure of 2,000 volts to twenty two arc lights mounted on poles along the three quarters of a mile of Broadway between Union Square and Herald Square.

Service was initiated on 18 December 1880. Madison Park was soon illuminated by lamps mounted upon a 160-foot tall mast that towered above the local buildings, the tallest of which were but fifty or sixty feet in height. Business establishments on 14th Street were offered service at the rate of ten dollars per lamp per month. Despite the relatively high price, the lamps were installed in Foster & Biel's Music Hall, the Park Theater, the Steinway emporium, the Gilsey House and a pair of hotels, the Stuyvesant and the Brunswick. Other businesses installed the lamps subsequently in order to remain competitive.

Brush had chosen his territory well. The popularity of arc lighting was such that additional stations were constructed to serve the commercial areas south of Union Square. Industrial plants, warehouses, and docks were the primary customers in the areas of lower Manhattan served by those stations. Brush soon had competition from Thomson-Houston and other companies. Electric street lighting was recognized by civic officials as a public betterment (not to mention a source of substantial graft) and charters were soon granted for additional installations. In 1887, eight companies received charters to illuminate the major thoroughfares of Manhattan from South Ferry to Coogan's Bluff at 155th Street. Many streets were illuminated only part of the night as lighting was reduced on some after midnight while lights on others were extinguished after that hour.

Indoor Electric Light

Thomas Edison was also determined to expand his market through the establishment of public service. The Edison Electric Illuminating Company of New York was incorporated 23 December 1880; five days after the Brush system lit Broadway. Although he was familiar

II: THE CENTRAL STATIONS

with the potential of the midtown area, Edison intended to construct at first only a demonstration project. For that he favored the downtown financial district in the belief that first-hand knowledge of indoor electric light would stimulate the interest of bankers that was vital to large-scale development of the system. Cognizant of the public objection to the overhead arc light, telephone, and telegraph lines already cluttering lower Manhattan, Edison insisted on the use of underground power distribution.

That noble intent conflicted with the "Gas Statutes" intended to limit the nuisance of street excavation through the restriction of permits to only those who planned to provide public service. Edison was thus forced to expand his program into a public system that covered an area of more than ten city blocks. The Edison Company was franchised to provide service throughout a territory bounded by Peck Slip, Ferry Street, Spruce Street, Wall Street, Nassau Street and the East River. That required generating capacity well in excess of any of those used previously. The result was the use of multiple machines, each rated at the electrical equivalent of 125 hp. It was also necessary to develop manufacturing facilities able to deliver the required components on schedule.

Those factories were located near the area to minimize the expense of transportation of the finished products. The generators were constructed by the Edison Machine Works that occupied the former Ashton Iron Works on

Goerick Street. The Edison Shafting Company was located on the same street while underground cable ducts were manufactured by the Edison Tube Company on Wooster Street. Electroliers (lighting fixtures,) fuses and other small components were supplied by the Bergmann Company founded by Sigmund Bergmann, a friend and former employee of Edison. It was located initially in the same area but success soon forced a move to larger quarters at Avenue B and East 17th Street. Thus Edison and his associates manufactured all of the electrical components; only the boilers, coal chutes, and steam engines were supplied by other companies.

Ironically, the capacity of the Edison lamp factory was inadequate to meet the demand and he was forced to underwrite the production of lamps at his own expense. In time, a large lamp manufacturing plant was established in Harrison, New Jersey to fill the need of the utility company. A power station of adequate size was provided by the purchase and reconstruction of adjoining buildings at 255 and 257 Pearl Street. Cable installation in the underground ducts was laborious and slow as the lines were insulated with pitch (tar) after the conductors were in place. Some problems developed; the most famous during the testing of the cables.

Strange reports of horses "dancing" on Fulton Street proved to be correct. Stray currents from failure of the insulation leaked to the street and reached the horses when their iron shoes came into contact with metal utility manholes in the street. The shocks were more

II: THE CENTRAL STATIONS

upsetting than serious as the incandescent lamps were designed for operation on 110 volt power. The same was not true of arc light systems. An electrical worker was electrocuted in a grisly spectacle on a line in front of City Hall. That primitive system lacked protective fuses or circuit breakers. The body remained on the pole, charring and smoking, until the current was disconnected when word of the incident reached the power station.

Much of the Edison generation and distribution system was the work of John W. Lieb, a twenty-two year-old electrician with a degree in mechanical engineering from Stevens Institute of Technology. Lieb had found employment initially with Brush but perceived better opportunity with Edison. Tests determined that the steam engine was the greatest source of potential problems. The first to be direct connected to the generator by a mechanical shaft (his small plants used belts) the engine produced excessive vibration and could not be "paralleled" with others to share the load. Attempts to do so resulted in uneven operation, as the engines "fought" one another, their speed and power output changed. Therefore, only part of the system could be energized at the official start of public service on Monday, 4 September 1882.

At 3pm, Lieb switched power into the distribution lines and Edison simultaneously turned on the lamps in the offices of the banking firm of Drexel, Morgan & Company. The system supplied four hundred lamps located primarily in banks and financial offices. Among the exceptions were the terminal of the stagecoach line that served Harlem, the office of the New York Herald, and Sweet's restaurant, a dining establishment once patronized by Thomas Jefferson and Aaron Burr. Though hailed in the press, the initial operation was difficult. The primitive generators proved cantankerous as the electrical contacts were subject to excessive sparking and burning. The coils developed excessive heat which required the installation of iron flues to improve air circulation. By November, new throttle controls eliminated most of the engine problems and all customers were supplied with electric service by the end of the year.

There were no meters to monitor voltage, only red and blue lights warned of over or under voltage conditions. Though primitive, that was an advance over early arc light stations. In one such plant, the only warning system was a length of pipe held on a field pole of the generator by the magnetic field. In the event the power system suffered a surge or failure, the pipe would fall to the floor with a loud clatter to alert the station operator.

At that point, service was supplied to nine hundred and forty-six customers in one hundred and seven buildings on Ann, Spruce, Fulton, Beekman, John, Pearl, Cedar, Pine, and Worth Streets. Lamps were available in two sizes, sixteen and eight candle power. The initial connected load consisted of 7,916 of the larger lamps and 6,395 of the smaller. The Pearl Street station had been preceded by an incandescent light service from the Holborn Viaduct station in London, England

II: THE CENTRAL STATIONS

the previous January. Development of that system was halted by political interference, and Pearl Street entered the history books as it continued to develop and expand without interruption after operation commenced.

Edison provided wiring, fixtures, and lamps at no charge. Replacement light bulbs were also provided at no charge, a factor that was no doubt conducive to the development of lamps of greater longevity over the next few years. The Edison Company introduced electric service at the rate of approximately 1.2 cents per lamp per hour, the first recorded instance in which customers were billed on the basis of power consumed. Usage was calculated by means of a "chemical meter" in which the passage of an electric current transferred metal from one plate to another by electrolytic action. The plates were removed and weighed at thirty day intervals to determine power consumption.

Electric lights became a status symbol desired by those who aspired to have their homes and offices graced with the finest appointments. Merchants soon discovered the merits of show windows illuminated by electric lamps, an innovation that changed the evening sidewalk scene forever. No business wanted to appear old fashioned at a time of fascination with new electrical developments. Gas lights or, even worse, oil lamps, marked a proprietor as being outdated if not somewhat eccentric. Edison had competition in the supply of electric service as Brush and others entered the incandescent light utility business. Private or "isolated" plants continued to supply many large institutions. The arc light companies continued to enjoy a large market in factories, stores and theaters as the available incandescent lamps were much too dim for large spaces.

System Limits

Incandescent electric light service met such rapid acceptance that additional capacity was soon required. However, the 110-volt system was an obstacle that imposed a one mile limit on the practical transmission range of a power station. Beyond that point, the internal resistance of the conductors produced a *voltage drop* (loss) that dimmed the lamps excessively. An immediate solution was promised by a "three wire" concept that connected two generators in series to double the transmission voltage while a balanced neutral wire between them provided the lamps with 110 volts. In that scheme, the loss was reduced to the extent that it enabled an effective transmission range of about three miles.

Capital expense of new installations was also reduced. As the voltage and current in a circuit vary in an inverse ratio, the doubled voltage reduced by half the current required to deliver the same amount of power. Since the current capacity of a conductor is determined by the cross sectional area, the reduction in current thus enabled the use of smaller wire. Invented in England by John Hopkinson, the 3-wire system was introduced to the Edison system in a manner which

II: THE CENTRAL STATIONS

led to significant future development. At the time Edison learned of the system, he was too occupied with other tasks to spare time to explore its potential value. He asked Edward Johnson to obtain information on the Hopkinson system. Johnson complied by contacting his friend Frank Julian Sprague who was then in London.

A Naval officer and recent graduate of the United States Naval Academy at Annapolis, Sprague had studied under the pioneer electrical researcher Moses G. Farmer at the Navy base at Newport, Rhode Island. That school was the leading electrical research facility in the world at the time and Sprague had been detailed by the Navy to report on an electrical exposition in London. He reviewed the Hopkinson "3-wire" concept in response to the request of Johnson. After returning to the United States and completing an extensive electrical text for the U.S. Navy, Sprague resigned his commission in order to develop the 3-wire system for Edison. Known as the "Edison 3-wire system" in the United States, it became the standard for the distribution of direct current service. Edison declared that the system reduced copper requirements by sixty-two percent.

As the area served by a power station tripled, the improved return on investment enabled more rapid expansion. Sprague also introduced mathematical modeling to the design of Edison systems. At that time, new systems were designed by the use of a crude empirical method that used miniature maps strung with fine wire. Energized by a small battery, the voltage loss across the map was measured with a sensitive meter and then extrapolated to the full scale installation. Fraught with error, it was only an approximation of service conditions. Astounded by the use of so primitive a design technique, Sprague set to work on the development of reliable standards. The rigorous mathematical formulae introduced by Sprague enabled successful installations on the first attempt.

Power Beyond Lighting

Another significant obstacle to return-on-investment was posed by the fact that electric light was used only part of the day. Edison noted that the financial district served by the Pearl Street station was a source of poor return as "banker's hours" did not extend into the evening when lights were most used. That was expensive, for Edison provided twenty-four hour service. Some of the power stations operated by the other companies were closed from dawn until late afternoon. In any case, the heavy investment in plant and system was used well only a few hours each day. Improvement of the *"load factor"* or utilization of the electric generating stations was vital if the industry was to survive and prosper. Industrial application of electric power offered the greatest promise for such growth.

At that time, industrial machinery was operated by stationary steam engines that were expensive, noisy, and

II: THE CENTRAL STATIONS

dirty. They occupied substantial space, required water, boilers, coal delivery, and ash removal. Power was transmitted to tools by a dangerous, inefficient, and maintenance intensive system of shafts, pulleys, and belts. Casualties among the workers were common when belts broke or connections failed. Many of those injured or killed were the women and young children who made up a majority of the factory workers of that period in history. Thus. factories and industrial plants posed a ready market for electric power, but no practical motor was available. Electrical literature abounded with schemes for an electric motor, but none were efficient. Most involved the application of power to a generator in order to force it to perform as a motor, a task for which it had not been designed.

The value of a practical motor had become obvious to Sprague when he supervised the installation of Edison systems in various cities and towns in the United States. He had also explored the practicality of an electric railway motor after a trip amid the smoke, soot, and cinders of the London Underground which was then powered by steam locomotives. Sprague decided to pursue the development of an industrial motor independently after Edison stated that he was interested primarily in light systems. With that decision, Sprague secured his future and changed industrial practice forever. Financed by Edward H. Johnson, he produced a practical industrial motor in time for the Philadelphia Electrical Exposition of September, 1884.

Powerful, it operated at a constant speed with little or no sparking, and it took industry by storm. Edison declared it the best available and the only one approved for use on his light system. Fans were recorded as the first motor load on the Edison system, a welcome innovation in the late summer and early fall of 1884. Other applications were found in short order. Overall, textile industries proved the most eager of the early customers and motor sales passed 3,000 in the first year. In one instance striking garment workers made the installation of electric motors one of the conditions for a return to work. As the Sprague motor eliminated steam engines from factories, it changed the appearance of industrial zones. Low factories were replaced by the loft buildings that came to characterize twentieth century industry.

Within a decade, forty-seven types of motors produced by Sprague and others found application on cranes, lifts, machine tools and other equipment. Worker safety improved as injury rates declined. Industrial productivity and the profits of factory operation increased. Electric utilities gained a substantial daytime load which brought phenomenal growth and financial stability. In 1890, the Edison Company of New York opened a "motor inspection bureau" to assist customers with the maintenance and repair of motors. By 1900, the motor load had increased by six hundred percent; most of that increase was that of large motors on printing presses and elevators.

II: THE CENTRAL STATIONS

Edison Leads the Industry

A number of factors established the Edison Company as the leading electric utility company in Manhattan. The innovations of Sprague, Lieb, and other inventors had produced a reliable system that provided customer satisfaction. The companies that Edison established to manufacture generators, lamps, and other components were merged to form the Edison United Manufacturing Companies. That guaranteed the operational compatibility of components. Most of his competitors had to depend on outside firms to supply vital components. The success of the Edison Company guaranteed access to capital for expansion. Finally, Edison had two significant talents. One was the ability to obtain talented assistants, inspire loyalty, and organize research effort. The other talent was public relations. That skill was demonstrated best by his steadfast opinion in regard to the superiority of underground service in urban areas.

After years of complaints about the lines that darkened the skies of downtown Manhattan, the Blizzard of '88 brought matters to a head. As snow accumulated on the lines, the cables and fasteners snapped and the wires fell to the street. Most systems lacked any protective device; the energized lines fell into the snow to create a hazard to man and horse alike. Overhead lines were outlawed in Manhattan soon after but many of the companies ignored the law. Some even challenged the ban in court. Months of frustration with the lack of compliance led Mayor Grant to direct the Commissioner of Public Works to enforce the law more directly. Ax wielding deputies were dispatched to cut live wires at both ends after which the dead sections were dropped to the street. A second crew then chopped the poles to the ground. With underground lines in place, Edison was spared the expense and delay entailed in line relocation. That made financial reserves and manpower available for expansion.

Edison stated that thirty six local power stations would be required to meet the demand for electric power in the area below 59th Street. Beyond the logistics of operation, the sheer capital expense would have doomed such a scheme before many were completed. Frank Sprague was retained to survey methods by which the situation might be improved. His report to the Edison Company of New York dated 13 September 1886 detailed current developments as well as future projections. He announced that in December he would have available a 220 volt motor that operated with greater efficiency and half the current of those in use. It would reduce the flicker of incandescent lamps and provide better service. That was the good news; the rest was not so simple.

Sprague had under development a pair of new motors. One operated on 400 volts for heavy industrial application. The other was designed for 550 volt operation to provide the power required to start and accelerate railway vehicles. Both required specialized power stations which represented heavy expense for the owners as well as lost business for the utility companies. Sprague

II: THE CENTRAL STATIONS

declared the 110/220 volt 3-wire system of generation and distribution both inflexible and inadequate for large scale systems. It was inefficient and expensive to construct separate power systems to supply arc lights, incandescent lamps, large industrial motors, and railways. Sprague favored the development of large power stations which could generate power in the amount required at less expense. He declared the use of high voltage the only practical means by which large amounts of power could be transmitted.

Experimental installations at the Newport Naval Base and at a mine in the western United States had convinced Sprague that the use of direct current at high voltage was impractical with the technology then available. The transmission of power to the mine from a distant power station had been effected by the use of 2,200 volt circuits. That voltage required the connection of machines in series across the line, a cumbersome arrangement that worked, but which constrained control and reliability. Sprague preferred the use of alternating current, which was still considered a laboratory curiosity by most experts. The primary argument in favor of alternating current was that it could be transmitted efficiently at high voltage over long distances. The voltage required by any particular load could be supplied by a simple transformer that did not require any moving parts. One generating station and transmission/distribution system could supply a variety of loads over a large area.

Beyond simple economy of scale, the use of alternating current promised flexibility in the size and location of urban power stations. Because range would no longer be a major concern, the plants could be located with respect to the electrical load center of a large region, the availability of land, and access to fuel. The transformer concept had been patented in the United States by William Stanley, an unlikely electrical pioneer. A law student from Brooklyn, Stanley had become so intrigued by electrical developments that he walked out of his law classes at Yale to pursue a future in the electric industry. In March of 1886, Stanley illuminated buildings in Great Barrington, Massachusetts with a 3,000 volt alternating current system that used transformers to supply the lamps.

Sprague cautioned that substantial research was required. Alternating current was not compatible with batteries. There was no practical alternating current motor yet available. As an interim step, Sprague proposed the use of inverted alternators (alternating current generators) as large "receiving" motors to drive direct current generators, both machines to be operated as a unit. They would be installed in "receiving stations" which would replace the local power station. The output would then supply customers over the 3-wire systems in the same manner as did the local direct current power stations. Thus the economy of scale inherent in large-scale generation would still be obtained

Sprague was committed too heavily in the development of railway and elevator motors to undertake new

II: THE CENTRAL STATIONS

research. Edison, committed heavily to the 3-wire direct current system in developed components, financial interest, and with installations around the nation and overseas, disagreed. He opposed development of alternating current systems even though Sprague had contributed much that had proven vital to the establishment of the Edison systems in place.

Edison In Midtown

Although the demand in the financial district was such that new facilities assisted the original Pearl Street station during times of peak load, Edison still considered the midtown area where Brush had installed his first arc light system to be the best potential market for nighttime lighting. Given the extent of the area to be supplied, the limited transmission range even with the 3-wire system, and the fact that efficient operation required that a power station be located at the electrical "load center" of the territory that it served, two stations were planned. The first structures planned as Edison power stations, one was constructed at 117-19 West 39th Street between Sixth and Seventh Avenues. The other was located at 47-49 West 26th Street, a short distance from the original Brush station. An arbitrary border was established on 31st Street. However, the street opening permits issued by the commission charged with the placement of power and communication lines underground were delayed for several years.

Edison Electric Illuminating Company
West 26th Street Station - 1888

II: THE CENTRAL STATIONS

The power stations were designated "vertical" stations; the components were stacked to reduce the requirement for expensive property in Manhattan. The engines were placed in the basement, the generators on the first floor. The second floor held the flues, ash removal equipment, and workshop. The boilers were placed on the third floor, with the coal storage bunkers, water tanks and meter room located on the fourth. The initial generating equipment consisted of two pairs of 100 Kilowatt (kW) belt driven generators, each powered by a 200 hp steam engine. Operation commenced at West 39th Street on 29 November 1888; West 26th Street followed on Christmas Eve. As events developed, the street opening permit issues delayed customer connections. The Union League Club which then stood at Fifth Avenue and 39th Street was the first Edison customer in midtown and was not connected until 27 December. Power from the West 26th Street station did not reach customers until 10 January, 1889.

Return-on-investment was substantial because new commercial buildings and department stores had been constructed in the west Twenties. New hotels and restaurants provided a demand for electric lighting into the late evening. Industries to the east and west contributed a substantial daytime motor load. Rapid commercial development in the upper Thirties enhanced the load factor of the West 39th Street station. Additional stations were opened to the north and south within five years, but the West 26th Street station remained primary.

Urban Life Transformed by Electric Light and Power

Beyond financial return and economic growth, the dawn of the "Electrical Age" brought social change. Electric light extended the hours available for amusement and entertainment. It gave new meaning to the term "night life." Electric signs soon adorned the theater district and Broadway which became known as "The Great White Way." The first such sign is said to have advertised "Heinz's 57 Varieties" on a building at Broadway and 23rd Street. The largest sign adorned the façade of the Cumberland Hotel which stood on the site now occupied by the Flatiron Building. It promoted the Manhattan Beach Hotel on the Atlantic Ocean in Brooklyn with the phrase "Swept by Ocean Breezes" spelled out in lights. The bright signs were not greeted with universal welcome. Proponents of city beautification had battled large billboard signs for years. Electric signs met with bitter denunciation and litigation. In time, the opponents lost and the art of electric signs advanced. The plain incandescent lamps of early years were soon replaced by colors and animated effects.

The residences of affluent New Yorkers included such innovations as electric carpet sweepers, vacuum cleaners, irons, cooking devices, and washing and sewing machines. An electric kitchen was installed at the Edison Company office in the Duane Street power station to provide demonstrations and free samples. At that time, utility companies were the primary source of electrical appliances. Only in the twentieth century did electric appliances become common in department, hardware, and other stores. The manufacturers of electrical products thus joined the utility companies in the promotion of electric service to architects, builders, and property owners. The companies collaborated on the production of an annual "Electric Show" that promoted new applications of electric power. Edison salesmen expanded their territory beyond homes and offices to market the service to large customers. Among the first were the Hippodrome Arena and the Metropolitan Opera House, both of which were only a short distance from the West 39th Street station.

Continued Expansion of Edison Territory

Despite the limitations of the direct-current system, the Edison Company supplied the majority of electric utility service in the area below 59th Street. Edison street lighting service was extended along Fifth Avenue to 72nd Street in 1892 as part of a pageant marking the 400th Anniversary of the landing of Columbus in the New World. Street illumination was still provided by arc lamps and a new design developed in 1889 operated on the standard Edison 3-wire system. By 1892, the company supplied 1,694 customers from a generating capacity adequate to illuminate 64,174 lamps of sixteen c.p. each. In modern terms, the total Edison capacity was about 3,600 kilowatts. Much of that power was supplied by the 26th Street and the 39th Street stations. The company ended the installation of wiring in 1892 to concentrate on electric service. Wiring was provided initially by the New York Electric Equipment Company Limited; soon after electrical "contracting" became a new business field for many laborers and professionals. The motor inspection bureau closed in 1897; its work was performed thereafter by repair companies that represented a new and expanding business in the electrical age.

The Edison Company established early a reputation for precise regulation of the power supplied to customers. Interruptions were rare; the only ones on record were a minor problem in 1883 and a major disruption on 2 January 1890 when the Pearl Street station was damaged extensively by fire. That reputation for reliability encouraged the use of the Edison system. Increased consumption was fostered through reductions in the price of power, the service was advertised at the cost per hour to operate a sixteen c.p. lamp. Service in the uptown district was initiated at 1.1 cents per lamp per hour, while the downtown rate remained the original 1.2 cents. Rates were standardized at 1.0 cents on 1 October, 1890. Two basic service classifications developed. The standard was the "Full Central" rate. The other was the "Breakdown" rate which was charged the operators of private plants that purchased power from the Edison Company only when their systems failed.

II: THE CENTRAL STATIONS

By 1900, the company supplied ten thousand customers at the cost of 20 cents per kilowatt hour. The kilowatt hour was the equivalent of 1,000 watts (one kilowatt) consumed steadily for one hour. That measurement appears to have been introduced with the advent of mechanical metering. Mechanical meters began to replace the old chemical meters in the mid- 1880s as the latter had been designed to measure by electrolysis the low current of lights. Meter readers with paper and pencil replaced the "meter men" who had driven carts through the streets to collect and replace the plates of the old chemical meters. An account of the period described piles of discarded plates from the old meters that were dumped behind the company offices as mechanical meters were installed. Improved fuses, switches and other protective equipment were also installed to handle the heavy current of motors.

Support by Battery

It had been established early that the peak load on the system always occurred in late December and early January when the short hours of daylight maximized the use of lighting in both homes and businesses. The development of Sprague electric motors for industries and elevators caused that trend to develop a "spike" (peak) between the hours of five and six in the evening. At that hour, the lighting load was heavy and elevators were in full use. As the load increased, large battery sets were installed to meet the peak demand without the need for additional expensive engines and generators which would have been idle much of the day. Instead, capacity which was idle at off-peak hours could be used to charge the batteries. Batteries also protected against system failure and became a staple of direct current systems.

Batteries were first installed in 1892 prior to installation of engines and generators at a new station on West 53rd Street that supplied the area near the northern fringe of the Edison service territory. The batteries were charged by generators at the West 39th Street station during periods of reduced load. The batteries were manufactured by Crompton-Howell of England and led to development of a new design for Edison stations. Batteries also enabled better seasonal use of generating capacity. In 1892, the West 26th Street station was the only twenty-four hour station in the system; the others were closed on summer Sundays when lighting loads were minimal and most mercantile establishments were closed. That did not last long; as load increased all the plants went to full time operation.

Elevators and Street Railways

Sprague motors changed the urban scene in the 1890s as much as they altered industrial practice in the 1880s. His elevators had an immediate impact on the city as they were installed in numerous structures. Furthermore, his patents were used by other companies that manufactured elevator components. Simple and efficient as compared

II: THE CENTRAL STATIONS

to the steam powered hydraulic elevator, his electric elevator made practical the "skyscrapers" that steel framing had made possible. Residential structures changed as luxury apartment buildings replaced the brownstone as an urban status symbol. In point of fact, the contributions of Sprague in regard to electric utilities, industrial motors, and urban architecture have been overlooked. While his later work earned him the status of "Father of Electric Railways," there has been little recognition of his prior achievements.

Unlike most cities in the United States, the electrification of Manhattan street railways was slow as a result of the substantial prior investment in cable powered railways. In time, the street car lines of Manhattan installed electric power, the cable vaults equipped with power rails because of the prohibition of overhead lines. Unlike many cities and towns, the transportation companies of the area constructed generating stations of their own instead of facilities shared with the local utilities. That was primarily a result of the density of traffic which required a capacity not available in the utility plants of the time.

Direct Current System Limitations Acknowledged

The Edison Company completed three direct current power stations after Edison left the active side of the business. The last of the Vertical style stations was constructed on Duane Street in 1891 to meet the demand of lower Manhattan. It made possible the retirement of the original Pearl Street station and small "annex" facilities which had been installed in downtown buildings. The aforementioned 53rd Street battery house and power station was opened the next year. A large station was constructed on East 12th Street in 1895 to supply the growing demand from Union Square theaters, restaurants and hotels and some east side industries such as dry docks of the marine trade. Unlike the vertical station, it was a large industrial building that presaged the appearance of 20th century power stations.

*Edison Electric Illuminating Company
East 12th Street Power Station*

II: THE CENTRAL STATIONS

Additional stations were deferred in view of the limitations imposed by the use of direct- current generation and distribution. It was obvious that a large-scale alternating-current generation and transmission system was required. That required time to design and capitalize, but the demand for electric power increased steadily. As an interim measure, the company continued to expand through the control or acquisition of smaller firms, most notably the Manhattan Electric Light Company which operated two stations on the East Side.

Improved efficiency was explored through such innovations as the DeLaval turbine imported from France. One was installed at the West 39th Station in 1896. The turbine provided a vast improvement in efficiency as it eliminated the heavy reciprocating mass of the steam engine and the complex system of valves, pistons and cranks. High pressure steam was directed against a series of fan like blades to produce smooth and efficient rotation. High-speed turbines operated at speeds of 750 r.p.m. or greater; the steam engine was limited to an average of 75 r.p.m. Although turbines required substantial development, it promised significant improvement in the efficiency of future power stations.

III: ALTERNATING CURRENT GENERATION AND TRANSMISSION

Frank Sprague may well have been the first electric system expert to recommend the total revision of electric utility practice in favor of alternating current. Sprague was associated closely with the Edison General Electric Company, the leading manufacturer of electrical components and an amalgamation of the Edison United Manufacturing Companies and the Sprague motor firm. The management was therefore familiar with his report. The company had noted European developments in alternating current systems, but had not licensed patents— allegedly because of Edison's opposition. As Sprague was occupied with elevators and railways, new pioneers emerged during the 1880s to make alternating current electrification practical.

Elihu Thomson patented an "induction coil" in 1879 which he developed into a practical transformer for an alternating-current arc light system that operated at a frequency of 125 Hertz (Hz) - cycles per second. One of those systems was installed in 1888 to power the portion of Williamsburg, Brooklyn that was supplied from the Penn Street Station of the Municipal Electric Light Company. That appears to have been the first alternating current utility in the New York City area. Thomson did not, however, pursue development of alternating current for residential or commercial customer applications because of a concern for safety. It has been alleged that he feared that a transformer failure might cause high voltage to enter the premises of a customer.

George Westinghouse thus took the lead in the development of alternating current for general application, especially in New York City. Westinghouse was respected throughout American industry for his railroad air brake. That invention, combined with subsequent signal system innovation, had increased the traffic capacity of the nation's railroads just in time to accommodate the demands of the new industrial age. He explored both gas and electric lighting and marketed a basic direct-current system in 1884. As a result of that experience, he set upon the development of a comprehensive system which could supply any load through a transformer.

Westinghouse declared the transformer a "most wonderful device" that operated silently by the transfer of energy between coils of wire through electromagnetic induction. He licensed the transformer patent held by William Stanley and financed that inventor's demonstration of alternating current lighting in Great Barrington, Massachusetts. Westinghouse licensed also

III: ALTERNATING CURRENT GENERATION AND TRANSMISSION

a number of significant European patents through the efforts of Guido Pantaleoni.

A student in physics at the University of Turin, Pantaleoni was familiar with the work of Professor Ferraris who is credited with having conceived— but never developed into practical devices—the alternating current theories conceived independently and patented by Nikola Tesla. According to the Westinghouse Company publication "The Electric Journal," the involvement of Pantaleoni came about through sheer serendipity. His father, Diomede, was a physician and personal friend of George Westinghouse. During a conversation with the elder Pantaleoni, Westinghouse extended an invitation to Guido to visit the Westinghouse plant in Pittsburgh, Pennsylvania.

Guido accepted and brought European developments to the attention of Westinghouse. Guido was then called home to Italy after the sudden death of Diomede. He subsequently became the European agent of Westinghouse and secured rights to the transformer patents of the English/French team of John Gibbs and Lucien Gaulard. On his return to the United States, he brought an alternator built by the Siemens Company of Germany. Pantaleoni subsequently entered the employ of the Westinghouse Company where he remained for five decades.

Like Brush and Edison, Westinghouse viewed New York City as vital to the successful promotion of his system. He entered the Manhattan arena in 1889 through the purchase of the United Electric Light & Power Company. Originally established as the United States Illuminating Company in 1880; the United Company was a small utility best known for having illuminated the Brooklyn Bridge with arc lights. The United Company introduced alternating-current incandescent lighting in lower Manhattan in 1889. The system was crude; alternators supplied power via a 1,000 volt line to transformers mounted on the walls of the premises of each customer. The fledgling effort enjoyed minimal success; the use of individual transformers was inefficient as the loss in each was cumulative. The losses in those primitive transformers were exacerbated by the use of a high frequency of 133 Hz. United supplied wiring and lamps at no charge in order to compete with Edison and the established companies. A few small midtown installations followed but major success was not forthcoming.

Undaunted, Westinghouse continued his efforts to improve the system and develop a practical alternating current motor through the acquisition of the patents held by more than a dozen inventors. The most important were forty patents that defined the polyphase (multiple circuit) system and induction motor of Nikola Tesla. Although aspects of the system were conceived independently by others, extensive litigation resulted in the decision that it was the Tesla patents that were primary to the large-scale application of alternating current. The Tesla system made possible a practical alternating current motor and addressed deficiencies in

III: ALTERNATING CURRENT GENERATION AND TRANSMISSION

the early transmission and distribution systems. His motor was based on a "rotating" magnetic field that transferred power by magnetic fields without cumbersome mechanical contacts.

Development of the various patents was directed by Benjamin G. Lamme, an engineer skilled in the conversion of abstract concepts into hardware that was viable in the commercial marketplace. Upon graduation from the engineering school of Ohio State University, Lamme sought a position with the Westinghouse Company when he heard about the new "Westinghouse Current" that was being developed in Pittsburgh. The term was noteworthy for it was illustrative of the degree to which the name of Westinghouse had become synonymous with the development and promotion of alternating current. Lamme determined that the polyphase system was ripe for immediate exploitation but that the Tesla induction motor required substantial development before it would be viable in the commercial marketplace. Lamme's sister Bertha Lamme, considered to be the first American woman to pursue an engineering field other than civil engineering, was employed by Westinghouse in the motor development program.

The Rotary Converter

The direct current motor designs of Sprague and others presented a formidable challenge to any alternating current motor design. They had undergone substantial development and many varied types and sizes were readily available. It could start heavy loads with ease and would accommodate easily short term overloads. It operated well at a wide variety of speeds. Speed control techniques were simple. It was the universal standard for railways, elevators, cranes, and other applications that required a motor to start a full load with precise control of speed.

Several types of alternating current motors were marketed in an effort to compete while development of the Tesla induction concept continued. Lamme developed for industrial use a variation of the Tesla motor known as the *"Wound Rotor"* type. Another, said to have been proposed by Dr. Louis Duncan of Johns Hopkins University, was a basic direct current motor modified for operation on alternating current. Known as the *"Universal"* type; it found a wide variety of applications. A third, first suggested by Elihu Thomson, was the *"Synchronous"* type, essentially an inverted alternator that rotated in exact synchronism with the alternations of the applied current.

As might be expected, the multiplicity of alternating current motors tended to confuse customers and reinforced the superiority of direct current machines and systems in the minds of many. While the aforementioned types of alternating current motors were applied to some machinery, direct current motors were more common. Therefore, any alternating current system designed for general applications had to be able to supply

III: ALTERNATING CURRENT GENERATION AND TRANSMISSION

customers with direct current. The link was provided by the *"rotary converter,"* a practical version of the receiving motor system that Sprague had proposed years earlier. Described by Lamme as "the overlaying of a synchronous motor on a direct current generator" the converter was less expensive than the use of an alternating current motor coupled to a direct current generator though that scheme was accepted in specialized applications. Lamme patented a rotary converter that enhanced the operation of alternating current systems as it balanced the load across the multiple circuits of a polyphase line. Small alternating current motors and those tested for railway applications tended to unbalance lines that were not designed specifically for those motors.

However, the rotary converter introduced a new complication. The ideal operating frequency was already a subject of considerable debate within the Westinghouse Company. Tesla proposed 60 Hz, a change which met with considerable opposition from Westinghouse engineers until it was ratified by the laboratory research of Oliver Shallenberger, one of Lamme's engineering staff. Rotary converters required a lower frequency for optimum efficiency and 25 Hz was adopted. However, the lower frequency was practical only for motors and converters as it produced a perceptible blinking of incandescent lamps known as "25 cycle flicker." The 60 Hz frequency preferred by Tesla became the standard wherever incandescent lights were operated on alternating current. Thus alternating current generation in the

United States was for many years a dual frequency system with 25 Hz supplied to railways and converters for direct current loads while 60 Hz was installed wherever lamps were supplied directly without conversion to direct current.

A Comprehensive System

Intensive development effort enabled the Westinghouse Electric and Manufacturing Company of East Pittsburgh, Pennsylvania, to market a comprehensive alternating current system in 1891. The comprehensive system was promoted as the means by which one power plant and transmission system could supply motors, incandescent lamps and arc light systems of any size. The Westinghouse polyphase system used two circuits, the second displaced ¼ cycle from the first. Most initial installations of the Westinghouse system were made in regions in which the transmission distance was too great for low voltage direct current. In other instances, the development of the area to be supplied was too sparse to amortize the expense entailed in the construction and operation of multiple direct-current facilities. In both instances, the initial response was enthusiastic and many experts believed that the Westinghouse system would soon dominate electric service. That confidence overlooked the heavy investment in direct-current systems on the part of urban utility companies and their customers.

III: ALTERNATING CURRENT GENERATION AND TRANSMISSION

Much has been made over the bitter debate between Westinghouse and Edison over the safety of high voltage alternating current. In reality, engineering practice was determined by the immutable laws of physics rather than heated emotional rhetoric. The Edison General Electric Company pursued intensively the development of alternating current as soon as Thomas Edison ended his direct involvement in the company. Intent on leadership in the electrical industry, Edison General Electric merged with the Thomson-Houston Company to form the General Electric Company in 1892. Promulgated by financier J.P. Morgan, the merger was to have included the Westinghouse Company. Adroit financial moves on the part of Westinghouse and his bankers allowed his company to remain independent. True competition developed as a result, a situation that was to accelerate considerably the evolution of electrical technology.

General Electric held the Thomson-Houston alternating current patents and it sought to develop a comprehensive system that would be competitive with that offered by Westinghouse. General Electric also purchased patents held by individual inventors and by small companies, the most significant of which were S-K-C and Siemens-Halske. The former, Stanley-Kelly-Chesney, had been founded by William Stanley after a short stint with Westinghouse. The latter was the U.S. affiliate of the Siemens Company of Germany. General Electric marketed a synchronous (rotary) converter through the patent of Charles Bradley, a young entrepreneur who had established the Bradley Electric Company in Yonkers, N.Y. Legend has it that Lamme believed his concept unique until he encountered Bradley in the United States Patent Office. In retrospect, the most important acquisition of General Electric was the Eickemeyer Company, a Yonkers N.Y. based firm that manufactured motors.

Rather than motor patents, the focus of that acquisition was said to have been the services of Charles Proteus Steinmetz. According to legend, so loyal to his employer was Steinmetz that he refused a General Electric offer until Eickemeyer was acquired. A brilliant mathematician, Steinmetz developed theorems that still define electrical engineering practice. In an effort to avoid possible infringement litigation with Westinghouse, General Electric had directed Steinmetz to develop an alternative to the two-phase system. His first effort was the "Monocyclic" system which consisted of a primary circuit and a secondary "Teaser" that provided the phase displacement necessary for the rotating magnetic field of the induction motor.

Monocyclic installations were limited and lacked the efficiency of the two-phase system. Steinmetz returned to the effort and explored a three-phase concept patented by Tesla and others but not yet developed by Westinghouse. That required three circuits, each displaced 1/3 cycle from the others. As developed by General Electric, the circuits were interconnected to permit the use of only three wires. The concept promised

III: ALTERNATING CURRENT GENERATION AND TRANSMISSION

an increase in capacity of fifty percent over that of an equivalent four-wire two-phase system. A long distance installation in Redlands, California initiated test operation about 1893.

However, at that time there was little knowledge as to a practical method by which single phase loads (lights and small motors) should be balanced on a three-phase line. Thus three-phase lines were generally limited to the transmission of power to three-phase industrial motors or to rotary converters for direct current loads. Alternating current customers were generally provided with power over two-phase systems, with the single-phase loads on one circuit, motors on both. Over time, systems evolved which provided single phase power from either circuit of a two-phase line.

Thus Westinghouse and General Electric drew even in the contest for dominance in the electric business. The two companies executed in 1896 a cross-licensing agreement which pooled patents to prevent the waste of resources through duplicative research efforts. That pool established the firms as the national leaders in the manufacture of electrical equipment and also as major contenders internationally. Thereafter, smaller firms tended to concentrate only on a portion of the market.

The Niagara Project

Westinghouse enjoyed an initial lead in the marketplace as the versatility of his system was demonstrated in a pair of significant applications. The first supplied hydroelectric power at 3,000 volts over a two-phase, four mile line to Tesla induction motors at a mine near Telluride, Colorado. The second connected a similar hydroelectric station to Portland, Oregon with a 4,000 volt, two-phase, twelve mile line. Those installations enabled the company to capture the contract to build the power station that supplied the 1893 Columbian Exposition in Chicago. Tesla and Westinghouse used that event to demonstrate the inherent efficiency and wide range of applications possible with alternating current systems.

That display changed the position of the Cataract Commission which had been appointed to develop an electric power station at Niagara Falls, N.Y. The Commission had favored direct current schemes but reversed its opinion in favor of Westinghouse after observing his systems in operation at the exposition. Contracts were issued and construction commenced by the end of 1893. The Niagara installation commenced operation in 1895 to provide the local area with two- phase power at 2,200 volts. A number of heavy industries relocated to the area near the falls and the impact of the project determined the future of the electrical industry.

General Electric obtained the contract to supply the twenty-two mile, three-phase line that delivered Niagara power to the city of Buffalo. The two-phase Westinghouse system at Niagara was linked to the three phase General Electric transmission line with an ingenious transformer developed by Westinghouse engineer

III: ALTERNATING CURRENT GENERATION AND TRANSMISSION

Charles Scott. Scott contributed much to the development of electrical technology and went on to head the Sheffield School of Electrical Engineering at Yale University. The line and connections were completed in 1896. The result established regional electric service and proved the claims made by Sprague a decade earlier.

Alternating current generated at the hydroelectric station at Niagara Falls supplied arc lights, railways, industries, commercial establishments, and residences. One station and transmission system replaced the seven separate direct-current generation and distribution systems that had been required previously. Simple transformers supplied alternating-current customers with the voltage required for their needs. Loads that required direct current were accommodated by *"substations"* in which transformers reduced the transmission voltage for the rotary converters that supplied the direct current for distribution to the customer. The universal impact of that project cannot be overstated.

Distribution and Customer System Considerations

Although the Niagara project had proven the superiority of alternating current for the generation and transmission of electric power, the system used for distribution of power to the customer was still a subject of much debate. Proponents noted that alternating-current distribution allowed a simple system with minimal components. The lines to the individual transformers were energized at a high voltage that reduced the current to the extent that only one-fifth the amount of copper was required to supply a given load as compared to an equivalent 3-wire direct current distribution system. However, serious concern persisted where alternating current distribution suffered losses in urban areas of dense load concentration. Some power was lost because of inductive effects in the cables but a larger concern was that of *power factor.*

Alternating current, when applied to the coils of motors and transformers, undergoes a loss of that power which sustains the magnetic field. It is termed *reactive* power as it is consumed by the inductive *reactance* in the coil. The lost portion supplies no usable power and is often called "wattless" power. It is also called a *lagging current* as the current wave lags behind that of the voltage wave as the cycle progresses. Reactive power is subtracted from the total *apparent* power (expressed as volt-amperes or va) to determine the *actual* or useful power (usually expressed as watts.) The term *"power factor"* was created to define the percentage of *actual* power of a system or component compared to the *apparent* power. Power factor concerns were addressed initially through the design of motors, transformers, and other components.

Transformers and motors perform best when operated at the power level for which they were designed. When operated at less than the rated power, the reactive

III: ALTERNATING CURRENT GENERATION AND TRANSMISSION

power consumed becomes a larger component of the apparent power and thus the percentage of actual power (power factor) is reduced. Power factor of the early designs in particular declined considerably when the units carried a light load. As recounted by Lamme in one of his many articles, research and experimentation with alternators at the Westinghouse plant led to the discovery that low power factor could be improved through the use of unloaded synchronous motors operated so as to create a *leading* current that would offset the lagging current. In that instance, the motors produced reactive power to compensate, or *correct,* power factor.

The notion that operating one motor without a load would offset the reactive power consumed by another that carried a heavy load appeared absurd to many. An early demonstration of the concept was that which accompanied the introduction of alternating current distribution by the Citizens Electric Illuminating Company of Brooklyn in 1895. That demonstration featured several small universal motors and a large induction motor. The total current was reduced by one third when an unloaded synchronous motor was switched into the circuit. Only those well versed in electrical engineering calculus could comprehend the facts. It was said that some observers refused to believe the indications of the meters! Synchronous motors that were intended for operation only to improve power factor were termed *synchronous condensers* and were vital to successful large scale application of alternating-current distribution.

Power factor regulation thus became a vital aspect of alternating-current system operation. Utility rate structures were developed to encourage the use of motors of good power factor and also the installation of compensation when beneficial. Still, it proved one more obstacle to the use of alternating current in the minds of many. Given the concern for inductive loss and power factor compensation, the majority of urban utility experts favored the retention of direct current distribution systems. As proven in Buffalo, small "sub" or "distributing" stations could replace the local generating plant in the supply of direct current to an area. In that approach, transmission power factor was contingent upon the design quality of the transformers and rotary converters, and also on the attention with which the converters were operated with respect to load. Direct current distribution also eliminated the concern for inductive losses in the high current cables that connected to the customers. To many experts it seemed the ideal solution.

Direct Current Selection

Steinmetz favored the use of direct-current distribution systems wherever practical and stated the succinct axiom that defined urban utility practice in the United States for more than two decades. That axiom based the selection on the density of load; direct current distribution was preferable as long as the customer base was

III: ALTERNATING CURRENT GENERATION AND TRANSMISSION

sufficient to amortize the expense of substations. With the exception of the United Company installations, direct current was the standard in the area of Manhattan below 135th Street that was supplied by the Edison Company or subsidiaries. At the time of the Steinmetz pronouncement in 1896, there was insufficient development in northern Manhattan to justify extension of the Edison lines. Direct current also remained the most common distribution system in the developed sections of Brooklyn as it was in the dense areas of most U.S. cities. Alternating current distribution was installed in other areas where load density was sparse.

Retention of direct current distribution proved attractive to both utility and customer. Electric companies found retention attractive as it required no change in the distribution system or meters and preserved the investment in battery reserve. Without batteries, the additional generating capacity required by peak loads would be superfluous most of the time. Furthermore, that retention expedited the substitution of a large, centralized plant for local power stations. The local stations could be replaced by substations gradually to reduce initial investment and minimize disturbance to operation. In some places, especially in Brooklyn, the transition from local power station to substation was expedited by the replacement of the boilers with transformers while the steam engines were replaced with synchronous motors. Those motors turned the original generators to provide power until rotary converters could be installed. Finally, retention assured future compatibility with potential future customers who then operated private plants.

Customers often favored the retention of direct current as no effort or expenditure was required on their part. A change to alternating current involved not only the purchase of new motors but also the practice of "overmotoring." That long forgotten term described the use of alternating current motors more powerful than the direct current motors they replaced in order to overcome the low starting torque of the alternating current motors available at that time. Overmotoring often led to the operation of motors below their rated load, a situation that reduced power factor further. Low power factor became a concern of customers once the utilities adjusted their rates to encourage efficient operation.

Direct Current Substations in Manhattan

The first direct-current substation in Manhattan was a temporary installation by the Manhattan Electric Light Company then under the ownership of the Edison Company. In 1896, a substation on Fifth Avenue at 72nd Street received alternating current from a generating plant on East End Avenue at 80th Street. It proved the concept, however widespread adoption of the scheme required a large central power station which had yet to be designed, capitalized, and built. The load on the Edison system had continued to increase as electric rates were reduced fifteen percent in an effort to capture the

III: ALTERNATING CURRENT GENERATION AND TRANSMISSION

business of the seven hundred and seventy-five private plants located in large office buildings, factories, hotels, and department stores. The success of that effort produced an immediate need for greater capacity. Short term relief was provided by a technique known as "double conversion."

Double conversion used an "inverted" (reversed) rotary converter to produce alternating current from the direct current generated in a local power station. That alternating current was then "stepped up" to a higher voltage for efficient transmission to another local station for conversion back to direct current for distribution to customers. The method had been pioneered in Chicago by utility magnate Samuel Insull to reduce the operation of many small stations that operated inefficiently with a poor load factor. Double conversion was installed in Brooklyn in the summer of 1897 to enable an Edison direct current power station in downtown Brooklyn to meet the summer demand of hotels and amusement parks at Coney Island. Double conversion entered midtown Manhattan in 1898 as the construction of a large alternating current power station awaited the rationalization and merger of a plethora of small companies.

That Manhattan double conversion linked the West 39th Street station with the Duane Street plant near City Hall in November, 1898. Each station was loaded heavily but the peaks occurred some thirty minutes apart. Termed the "Broadway Cable," the link enabled one to support the other, effectively buying time until a large station could be completed. It was extended subsequently to the 12th Street station in anticipation of construction of a large central power station. It also required a new level of control. Small local power stations were distinct entities supervised by the engineer in charge until the linking of the stations necessitated an overall supervision of the system. The office of the *"Load Dispatcher"* was established at the Duane Street station in 1898 to monitor the network and assure the most efficient and reliable utilization of available capacity. That office marked the transition of Edison service from autonomous local power stations to an interconnected system.

Railway Systems

As the rationalization of the small utility companies of Manhattan commenced, the first large alternating-current generation and transmission systems in Manhattan were constructed to supply railway service. Those companies were burdened by the same concerns that plagued the lighting companies. Success had produced growth and expansion on a level that required additional local generating stations. Some were constructed, but the capital and operating cost of multiple direct-current power stations was considerable. A single large-scale alternating-current power station with transmission to substations promised to remove the limitations imposed by the small plants. The Metropolitan Street Railway

III: ALTERNATING CURRENT GENERATION AND TRANSMISSION

opened in 1898 an alternating-current station on the East River at 96th Street to supply the entire streetcar system.. Three-phase, 25 Hz power was transmitted at 6,600 volts to seven substations located along the lines. Typical of the systems installed by street railways across the nation, that installation was eclipsed subsequently by the gargantuan undertaking of the Manhattan Railway Company.

In 1898, the elevated railroad lines operated by the Manhattan Railway Company were vital to the commerce of the city. More than two hundred million fares were collected annually as New Yorkers of all economic classes depended upon the system for access to employment. Capacity was constrained by the limited power of the steam locomotives of the time. The company experimented with various schemes to improve the motive power, but none proved practical. Initial electric power schemes were thwarted by the limited capacity and the expense of small power stations. The propulsion system required a revision, for the simple replacement of steam locomotives with electric units would not have achieved the goal of increased train length to accommodate new traffic. To do so would have required heavier locomotives as the pulling force exerted by a locomotive is a function of the total weight on the driving wheels. However, the locomotives in use were close to the weight limit imposed by the weight-bearing ability of the elevated structures. Two major developments resolved the issues in 1897.

First, an examination of the performance of the Niagara Falls power system after a year of operation indicated that large-scale generation and transmission could supply sufficient power.

Second, the locomotive issue was eliminated by the *Multiple Unit* control system demonstrated by Frank Sprague on an elevated line in Chicago. In that concept, each car was equipped with motors and controls, the latter linked by a *train line* cable to a master controller on the first car. Train length could be varied to meet the dictates of traffic as the weight of each motorized car was within the limits imposed by the design of the elevated structures. The Manhattan Company initiated an eighteen million dollar electrification program in January, 1898 upon the condition that it was directed by Lewis B. Stilwell. He had directed the Niagara project, and the Manhattan Company deemed him to be the best qualified to direct the largest electrification project of that time. The alternators were rated 5,000 kW, the largest ever attempted.

Westinghouse supplied the generation and transmission system. The power station on the East River at 74th Street produced three-phase, 25 Hz power at 11,000 volts for transmission to seven substations along the lines. Westinghouse rotary converters in those substations produced 625 volt direct current distributed by third rail to the trains. The substations also supplied power for signals, shop buildings, and the illumination of yards and passenger stations. Success of the effort was

III: ALTERNATING CURRENT GENERATION AND TRANSMISSION

manifest from the first day of public operation in 1902. Elimination of the steam engines increased capacity of the lines; traffic increased and the investment was recovered in a short time. While the concept had been validated previously by smaller installations in Chicago and in Brooklyn, the size of the Manhattan installation surpassed any previous effort. The layout and operation of the system was studied by engineers from electric companies, utilities, and railroads across the nation and around the world.

A revolution in motive power practice followed the success of the Manhattan elevated railway installation. A virtual duplicate powered the subway lines that opened in the new century. The New York Central, Pennsylvania, Long Island, and Hudson & Manhattan railroads installed similar systems within a few years. New York City became the world capital of electric transportation with a network that eventually included streetcar and rapid transit lines in every borough as well as mainline railroads that entered the city from the north, east and west. With the exception of freight on the west side, Manhattan achieved smoke-free railroads within a decade. By 1910, a total of ten large alternating-current power stations had been constructed to supply power to substations of the region's railroad and transit lines.

IV: GROWTH THROUGH ECONOMY OF SCALE

The Better Future

At the turn of the century, electrical science was viewed as an important development that promised better living in a better society. Electric power systems were already vital to modern living and a potential source of income to laborer and investment banker alike. Developments in electrical technologies were the subject of feature stories. Even the Stratemeyer Syndicate, which later produced such juvenile classics as the Hardy Boys stories, had found the electric industry useful for the background of a story. Stratemeyer released in 1897 a volume entitled "Bound to Be an Electrician," the tale of an ambitious youth in the electric business.

The new century brought new technical development almost weekly. New "electrical steel" was formulated with magnetic characteristics that improved the efficiency of alternators and transformers. The capacity of alternators and rotary converters doubled every few years. The adoption of steam turbines was the greatest improvement in the power stations constructed in the new century. The first turbines built in the United States produced twice the power of the steam engines they replaced. Further improvement increased their output by a factor of six over the first decade. As those occupied the same space as the original steam engines, they increased the capacity of the stations in which they were installed by six hundred to twelve hundred percent. Metal filament "Gem" (General Electric Metalized) lamps were developed. Tungsten filaments subsequently doubled the light output per watt of power consumed.

As economy of scale reduced electric rates, the consumption of power increased. Advertising for new apartments noted prominently "Electric Lights." Utility advertisements exhorted the owners of commercial as well as residential properties to install electric service to increase the value of their premises. Increased sales of appliances produced lower prices as the average citizen found that small appliances were affordable in both first cost and power consumption. Specialized retail outlets and repair shops added to the economic growth. Within a decade, novelties such as electric toys and Christmas tree light sets were no longer restricted to the very wealthy.

Architect of Power

The reorganization of Manhattan electric utilities was directed and funded by Anthony N. Brady, an executive

IV: GROWTH THROUGH ECONOMY OF SCALE

experienced in the gas and street railway companies of Albany, N.Y. Brady set about acquiring the numerous and sometimes conflicting electric utility franchises in New York City. His point man was Thomas E. Murray, an engineer also from Albany. Murray entered the utility business of New York City as the General Manager of the Central Telegraph and Electrical Subway Company; the firm that constructed and maintained the underground ducts used by the electric utilities. A gifted engineer, inventor, and manager, Murray consolidated small utilities into substantial companies to raise the capital required to expand electric capacity. The phenomenal growth of electric service in New York City after 1900 was attributed largely to the effort of Murray. A leader in technical innovation, Murray designed the power stations that supplied New York City utilities for the first half of the twentieth century. His first station in the city opened in 1900 on Gold Street at the Brooklyn waterfront near the Navy Yard. That was followed by a plant for the Brooklyn Rapid Transit system in 1901. After the reorganization of the electric utilities in Brooklyn, Murray brought order to the electric companies of Manhattan in 1901.

The capital of the Consolidated Gas Company of New York was the primary tool of the reorganization program. About the time that Edison left the electric utility business in 1889, the gas company purchased control of the Edison Company of New York. Brady acquired seven small Manhattan companies in 1900, as the Edison Company merged with two others. The Brady holdings were merged with the Edison Company in 1901 to form the New York Edison Company. The United Electric Light & Power Company was acquired by Consolidated Gas in 1900 but continued to operate as an independent entity. It was maintained as a parallel operation which retained a separate corporate and operational structure because of the differences in the power systems. United supplied 60 Hz alternating current as the standard while New York Edison installed 25 Hz generation and transmission with substations that supplied direct current to customers in Manhattan.

Although the companies were in competition so far as the public was concerned, they cooperated in technical matters. Annual reports to stockholders were filed jointly as units of the Consolidated Gas Company. The large majority of New York Edison customers were located in the area south of Central Park which continued to lead the world in development of electric service in urban areas. Several years were to elapse before United secured a significant commercial and residential load.

The overall Brady/Murray concept defined both the Edison Company of New York (Manhattan) and that of Brooklyn plus the United Company as the "Generating Companies" which had a sufficient customer base and potential income to capitalize the construction of large scale generation and transmission systems. Smaller companies without such potential customer base were located in Westchester, Queens and the Bronx. Those were termed "Affiliated Companies" which would be supplied with power by the Generating Companies. As such they would benefit by elimination of the small and inefficient plants that were in operation at that time. The generating companies would also benefit by an increased market and greater economy of scale.

Waterside

The design of a large generating station to supply Manhattan was initiated by the New York Gas and Electric Light Heat & Power Company in concert with the Edison Company. The site chosen on the East River at 38th Street provided water access for coal delivery and ash removal. Unsubstantiated mention in at least one record indicated that a small local direct current power station may have operated there from 1899 to 1909 but no details were provided. The location also represented a midpoint between the northern and southern limits of the Edison service territory, ideal for efficient power transmission from a large station. Named Waterside, the plant was to have featured an ornate facade that included offices and apartment towers for the operating crews. Brady advanced the Waterside plan in 1899 and placed it under the direction of Murray. Murray incorporated his experience with the Brooklyn stations into the design of Waterside and eliminated the towers and residences.

Waterside captured the very essence of the new century, a time of optimism and confidence in the future. Immense by the standards of the day, it occupied an area of 272.5 feet by 197.5 feet on a foundation of concrete and bedrock. Massive conveyors took coal from river barges to the storage bins that held it for transfer to the furnaces. Ash conveyors and condensing water tunnels passed under a marginal street that bordered the river. Within the plant, two floors on the north side held rows of fourteen water tube boilers, twenty-eight to a floor, fifty-six in all. Steam was delivered by ten-inch pipes to the engines that drove the alternators.

Planned to accommodate either engines or turbines, the station was equipped initially with the proven technology of reciprocating steam engines. Triple-crank marine engines with three cylinders, one high pressure, the other two low, revolved the alternators at 75 r.p.m. The alternators were each rated at 3,500 kilowatts, the capacity of the entire Edison system a decade previously. Eleven units were installed before it was decided to adopt turbines. The first, a 5,000 kW turbine alternator, was placed in service on 2 September 1904. It was said that Murray funded the generator out of his own capital because Brady refused to allow the New York Edison Co. to take the risk of investment in a new technology at such a critical location. The last unit was rated 8,000 kW.

The station operator was linked to other facilities by telephone lines and a Gamewell alarm system similar

IV: GROWTH THROUGH ECONOMY OF SCALE

to those in use at that time by electric railways and fire departments. Special precautions were taken to guard against fires throughout the station, particularly in the coal storage areas. Operation of the entire system was monitored and directed by a "system operator" that replaced the Duane Street Load Dispatcher in 1902.

On 24 October 1901, the first Waterside power was sold to the public. Underground cables carried three-phase, 25 Hz alternating current at 6,600 volts to the substations that provided the direct current to customers. The Broadway Cable that linked the 39th Street station with the Duane Street and 12th Street stations became part of the new transmission system that delivered Waterside power to eight substations which had been constructed as part of the Broadway Cable program in anticipation of the Waterside station. New substations opened following the completion of Waterside, and all of the local Edison direct current stations were equipped with transformers and rotary converters to initiate their transition into substations. In time, the generating equipment in those local stations was retired. An initial equipment order totaled forty-four rotary converters from the S-K-C Company. The largest were installed in the major substations; four units rated at 2,000 Kw, and ten rated at 1,000 Kw. Thirty small units rated at 400 Kw. each were installed throughout the system. By 1903, the capacity of the Edison system was 50,600 Kw, more than a ten-fold increase in as many years.

*New York Edison Company Waterside
Stations #1 (1901) and #2 (1906)*

*NY Edison Co. Waterside Station #1 "Operating
Room" where power was generated*

IV: GROWTH THROUGH ECONOMY OF SCALE

*NY Edison Co. Waterside Station Control Board
for outgoing power feeder cables to substations*

Aesthetics

As the primary supply of electric power in Manhattan, the Waterside station was the subject of articles, books, photographs, and renderings in oil and charcoal. At any hour of the day or night, the hum of machinery could be discerned from the street. It remained a symbol of American technical achievement for decades. New developments were reported regularly and Waterside became the subject of visits and compliments from electrical power experts of nations throughout the world.

Even those without an understanding of electrical science came to equate the name "Waterside" with electric power, much as the term "Niagara Falls" held the same implication a few years previous.

Neighborhoods were improved as Waterside brought about the end of the local power station and the attendant noise, coal smoke, and clouds of waste steam. Fuel delivery and ash removal no longer blocked sidewalks and interrupted street traffic. The retirement of private plants reduced further that clutter and pollution as owners substituted the low-cost utility service made possible by the efficiency of Waterside. Although the plants disappeared, the substations remained in the neighborhoods as symbols of the electrical age, and were given a variety of eclectic architectural treatments to assure visual harmony with adjoining buildings. Most had the appearance of commercial buildings without any visible commerce.

If a passerby happened to observe the interior of a substation through an open door or hatch, all that was to be seen were ranks of large machines emitting a steady hum as the armatures rotated to convert the power. Otherwise, they were nearly silent, the company having expended considerable effort to isolate the operating components from the structures to minimize the transmission of vibration. Most continued the vertical format to fit one or two of the narrow lots of Manhattan. The converters and transformers were installed in rows, the auxiliary controls and related equipment were placed on

IV: GROWTH THROUGH ECONOMY OF SCALE

upper floors or mezzanines. Later extension structures followed the same approach in most instances.

The West 39th Street Complex

In 1903, substation equipment replaced that of the 1888 vintage generating station. It supplied an area in which the construction of large hotels and restaurants followed the establishment of Times Square as the theatrical center of the city. The territory supplied by the West 39th Street substation was divided at Madison Avenue in November, 1906 when a new substation opened on East 39th Street between Park and Lexington Avenues. (Records indicate that a substation was added at 32nd Street and Lexington Avenue in January of that year but no further details were given. It apparently was in operation for only twenty-seven months until March of 1908.) The East 39th Street substation carried primarily residential load but also some significant commercial customers that included the Grand Central Palace exposition center and fashionable restaurants such as Sherry's and Delmonico's.

Still, the load of the Times Square area continued to expand at a rate which required an extension of the West 39th Street substation in 1908 and two years later, the addition of the "Gimbel's" substation. That was located three levels below the main floor of the large Gimbel's department store under construction on Sixth Avenue between 32nd and 33rd Streets. Most major department stores in the midtown area such as B. Altman's, Macy's, and Sak's operated their own private plants. The Gimbel's substation supplied both the store and the surrounding neighborhood from five 1,000 kW converters. That fact was quoted often by Edison salesmen in their efforts to convince property owners to replace private plants with Edison service. Rare in New York, such "basement" substations were said to have been common in Chicago.

An additional substation was constructed on West 41st Street off Eighth Avenue to assume the load of the new offices of the New York Times, the Paramount Building, and some large hotels. Still, the West 39th Street substation carried a heavy nighttime load from the theaters, restaurants and hotels of Times Square and large lighted signs. The most famous of the latter was the "Roman Chariot Race" sign on Broadway at 38th Street. The twenty-four hour load required such a large staff that the substation was selected as the location for the company Emergency Department which responded to breakdowns. The original structure of the 1888 power station was replaced in its entirety with a new and more spacious structure in September 1915.

The West 26th Street Complex

The electrical load of the midtown area that was once designated the Edison Company's "Second District" increased so rapidly that substation capacity had to

IV: GROWTH THROUGH ECONOMY OF SCALE

be increased at frequent intervals. The first substation in that area was constructed adjacent to the 1888 26th Street vertical station when Waterside opened. It was supplemented by an extension in 1904 that included a library, lecture hall and training school for employees. An additional substation was constructed on West 27th Street near Tenth Avenue; three years later, another was located on West 16th Street near Sixth Avenue. While the territory supplied by the West 26th Street complex had been reduced, the load trebled because of the construction of new loft buildings and large commercial structures. In 1909, the original power station was rebuilt as a substation so thoroughly that a company history of 1912 stated that little more than the arch over the door remained of the 1888 structure. By that time, the territory supplied by the station had been reduced to accommodate the increased load of the area. By 1912, it was rated at 18,500 kW with eleven rotary converters in the three structures.

Madison Avenue became the eastern boundary for the West 26th Street substation as well as for the West 39th Street substation when customers east of Madison Avenue were transferred to a new facility on East 26th Street near Third Avenue in 1913. The largest single load transferred at that time was the Metropolitan Life Building at East 24th Street and Madison Avenue which had been the tallest building in the world when it was completed in 1909. An extension was to the West 27th Street substation was added in 1919.

Still, the West 26th Street substation remained the most heavily loaded of those operated by the Edison Company at that time. It was likely the most heavily-loaded utility substation anywhere in the world. It shared with the West 16th Street substation the load of large department stores along Broadway and Sixth Avenue. That territory included also many loft buildings with a large number of industrial motors. The heavy motor load in those factories often caused sudden power surges. Precise regulation of voltage was important because merchants on the "Ladies Mile" objected vociferously to any flickering of the lights in their fashionable mercantile emporia. Many of them retained private plants that could be restarted in the event of dissatisfaction with Edison service. Those stores moved uptown over time but the electrical demand increased as new industrial tenants occupied the old buildings.

Substation Operation

Within the substations, transformers reduced the 6,600 volt alternating current from Waterside to 170 volts for the rotary converters. The direct current from the converters was then supplied to outgoing *feeders* and also to the batteries of the substation. The connection to the batteries was selective, the batteries were charged when the load carried by the substation was low, provided adequate capacity was available at Waterside. The batteries were then used for support

IV: GROWTH THROUGH ECONOMY OF SCALE

when the load on the substation and/or Waterside was heavy. The transmission range of the substations averaged 5000 feet, the feeders connected at underground junctions to form networks.

From those junctions a set of *mains* provided power to the premises of each customer. Direct current was supplied at 120 volts for residential and light commercial needs, 240 volts for heavy commercial and industrial loads. The company took pride in the precise regulation of voltage. Good regulation of voltage assured proper performance of motors, bright lights, long lamp life, and customer satisfaction. It was also a selling point in comparison to private plants which were often unstable when loads were heavy or subject to sudden change. An ingenious "pressure wire" monitored the end of the feeders to facilitate voltage regulation.

Indications at the substation enabled the skilled operators to switch the outgoing feeders as necessary. Three voltages were available at the converters for the outgoing lines: 123, 128, or 133 volts. The selection was determined by the "end pressure" indicated by the pressure wire. In that way, steady voltage was maintained during times of increased load. Prior to the development of automatic regulation of the rotary converters in the 1920s, the voltage supplied to customers was dependent on the skill of the substation operator. In general they were excellent, the switching of the feeders detectable only by occasional minor changes in the brightness of lamps.

NY Edison Co. West 16th Street Substation

IV: GROWTH THROUGH ECONOMY OF SCALE

NY Edison Co. W. 16th Street substation interior showing unusual vertical shaft rotary converters

Direct Current Distribution Expert

Provision of reliable power in the face of such change and development required not only accurate forecasts but also precise work schedules to coordinate the frequent modification of substations, feeders, and mains. The substation and distribution system was developed and expanded under the direction of John W. Lieb, the man who had designed the generators for the Pearl Street station in 1882. In subsequent years, he had directed the installation of Edison lighting systems overseas. The system he installed in Milan in 1883 was considered by Edison to be the best in Europe.

His success in Milan earned him a decoration by the Italian government in later years. While working in Milan, he uncovered in excavations some remnants of works dating to Leonardo DaVinci, a discovery which led to a lifelong fascination with that artist and inventor. Lieb assembled the best collection of publications on DaVinci to be found in the United States, an assemblage that became the keynote of subsequent exhibits. Lieb had returned to New York in 1894 to direct the expansion and development of direct current distribution systems.

Selling the Service

New York Edison expanded the intensive marketing established decades earlier. Sales bulletins were frequent, newspaper and magazine advertisements were constant. A cartoon character named Father Knickerbocker was created in 1905 by F.G. Cooper, an artist destined for fame in later years. Dressed as an eager workman rushing to do a job, he carried a loop of wire and huge toolbox. In time, Father Knickerbocker became as familiar to New Yorkers as Uncle Sam; his image adorned company stationery, buildings, vehicles, and signs for sixty years. The company slogan was "At Your Service." Two magazines were published: the *Edison Weekly* for employees, and the *Edison Bulletin* for the public.

The latter evolved into the *Edison Monthly*, a glossy magazine about New York. Additional publications

IV: GROWTH THROUGH ECONOMY OF SCALE

were produced for major events. Others depicted the city at night, or scouting and civic activities sponsored by the company. Books were published to mark historical anniversaries, detail technical developments, and depict the services provided. The sales force promoted the simplicity and economy of Edison power in comparison to the expense entailed in the construction and operation of private plants.

Lighted signs were used extensively to promote the services. Large signs adorned power stations and offices. Attractive lighted window displays were placed in the general office at 55 Duane Street and also in the district offices. One large illuminated sign proclaimed that the new building of the *New York Times* was supplied with Edison service. Another changed sequentially to read "LIGHTS," "POWER," "ELEVATORS." In 1906, the company operated sixteen large signs that used a total of 10,654 lamps. Three years later, the company participated in the Hudson-Fulton celebration by outlining the buildings at Waterside with 30,000 lights. The smokestacks were wrapped with a spiral ring of lights. The event marked the anniversaries of the 1609 navigation of the Hudson River by Hendrick Hudson and the 1807 demonstration of the steamboat by Robert Fulton.

That celebration took the grandeur of outdoor lighting to a new level with 30,000 incandescent lights as the celebrations opened on 25 September 1909. The harbor was the site of a massive nine mile display of ships; fifty five of them from the U.S. Navy, the balance from eight other nations. As celebrations were held in cities and towns along the Hudson River, replicas of historic vessels circled the parade of ships. Monuments on the shore were lighted with strings of lights, the Soldiers and Sailors monument at 96th Street and Riverside Drive was strung with 1,500 lamps. Ryan "Scintillator" Searchlights were mounted at locations on high ground to project beams skyward. The East River bridges were strung with lights.

The Washington Square arch was outlined with 1,500 lights while City Hall was outlined with 3,000 lamps. The arches of the Riverside Drive Viaduct over the Manhattan Valley at 125th Street were outlined in light as was the Bronx Civic Center. The new Public Library site at 42nd Street and Fifth Avenue was made the centerpiece of a "Court of Honor" with thirty-six stucco columns fifty-five feet high that were placed along the avenue and strung with garlands and lights. Buildings on the Jersey shore were also illuminated. Lights were strung along other parts of Fifth Avenue. Major streets were illuminated with lamp strings while the facades of hotels, department store, and other commercial structures were lighted. The entire structure of the new Metropolitan Life Building on East 24th Street was illuminated with a variety of spotlights and floodlights, as well as strings of lights. A streamer of light was suspended between the spires of St. Patrick's Cathedral. The celebration lasted for two weeks into early October and it was noted at the time that an equal display would probably never be seen again. Temporary circuits supplied power from the United and NY Edison systems while private plants contributed to the illumination of the buildings in which they were installed.

A smaller version of that celebration followed on 17 January 1912 when temporary lines were run from the 53rd Street Edison substation to outline St. Patrick's Cathedral with 50,000 lamps with 20 miles of wire. The 340' spires were illuminated with 11,000 lamps, the steeplejacks who performed the work had to be specially trained in electrical wiring. All the workers were provided with electrically heated coats as work was impeded by a biting wind which was followed by snow that turned to rain and then froze at night to create a severe ice hazard. Twenty-five men were assigned to maintain the system and replace failed bulbs. Designed by the architects Lamb and Poole, the display marked the new lighting of the Cathedral which was one of the largest interior spaces to be equipped with new lights. The event was scheduled to greet the return of Archbishop Farley from Rome and then extended for a few days. Like the Hudson- Fulton display, it was a display of electric lighting that would never be forgotten by those who witnessed it.

Lamp Supply

Although motors consumed most of the power sold, the company continued to supply free light bulbs in the early years of the new century. When an account was opened, a company representative would deliver an amount of bulbs sufficient for the premises and about twenty percent extra to be held by the customer as spares. Burned out bulbs were exchanged by the delivery man or at company offices. Special "Hylo" or "turn down" lamps that could be dimmed were sold without free replacement. The exchange of light bulbs was a significant activity in itself. Twelve small electric wagons, each stocked with one thousand to three thousand lamps, circulated throughout the Edison franchise area. The stock was replenished by a heavy electric truck which made the trip to the General Electric lamp factory in Harrison N.J. an average of five times a week. Each trip delivered about 45,000 bulbs. In time, the company was forced to install lamp crushing machinery at the Waterside station to dispose of the burnt out bulbs that were returned. The provision of free lamps ceased about 1915.

Company Morale

New York Edison sponsored a variety of programs and events for employees. Beyond the usual dinners, picnics, and bonuses, the company encouraged staff members to read and learn about their work. The company gained a better-informed staff; the employee achieved promotion and earned better income. In an era when many young men went to work at fourteen years of age or younger, New York Edison sponsored extensive activities in concert with the Boy Scouts of America. Company publications gave extensive coverage to hikes and camping trips in all seasons. Company efforts aside, the workers took pride in being on the threshold of a new industry. Generating station and substation crews felt a special satisfaction on passing a humming factory or seeing lighted buildings at night knowing their labor made it possible. Many related taking their families to roof tops, elevated railway stations, or other high locations at night to see the city lights.

New Demand

The annual electric show by manufacturers and utilities encouraged the use of electrical devices. The utility companies mailed a pair of free tickets to each customer; the expense was exceeded by the profit from increased sales. Industrial and domestic machines were promoted with exhibits and displays that emphasized action. Electric vehicles, especially trucks, were promoted as the company sought the business of charging the batteries of electric vehicles at substations. Prior to the development of gasoline motor trucks during the 1920s, electric trucks were a staple of New York businesses from bakeries to department stores.

The company operated a fleet of thirty-seven electric trucks in 1906. Seventeen were one thousand pound-capacity electric wagons used for the delivery of lamps and appliances. Seventeen were two thousand pound-capacity emergency service wagons. Three trucks of six thousand pound capacity performed major tasks such as the drawing of heavy cable through underground ducts. Truck repair and storage was located at Waterside. In an emergency, with authorization from Supervision, the company trucks could be recharged at street lamps. At that time, the New York Edison Company provided the majority of the lights in public areas.

In addition to the commercial and industrial applications of electric power, new domestic appliances were promoted vigorously. Attractive displays of home appliances were placed in company offices. In most homes, the appliances were limited to irons, heaters, fans, and toasters but an extensive array of appliances was available. Specialized appliance shops opened to serve that market and electrical sections were added to hardware and supply stores. Routine repair of appliances produced dozens of new "electric shops" throughout the city. Some were affiliated with merchants, others conducted business independently.

Electrical World detailed the range of those items in the description of an exhibit in the fall of 1906. Held by Wanamaker's Department Store, it featured the items available for the well- appointed home. Wanamaker's was entering a new market, as many potential customers preferred to purchase appliances from electric utility companies. Most felt that they could depend upon the warranty repair provided by the electric company, whereas the department stores lacked experience in that

IV: GROWTH THROUGH ECONOMY OF SCALE

area. Wanamaker's acknowledged those concerns with a plan to provide warranty repairs. Like most electric exhibits at that time, the rooms of a well-furnished home were considered the best showcase for domestic appliances.

The kitchen was most prominent, dominated by a massive refrigeration plant that bore little resemblance to the refrigerators of later years. Electric ranges and ovens were accompanied by electric plate warmers and even a dish pan that was heated electrically. Food preparation equipment included a coffee urn, coffee grinder, teapot, and meat and vegetable grinders. Good care of garments was assisted by washing and wringing machines, irons, and sewing machines. Household chores were simplified by carpet sweepers and a polisher for silverware. Personal comfort was provided by fans, heaters, heating pads, and curling irons for the hair. While it may be stated with certainty that such extensive use of electric appliances was rare, the development of that market did constitute a new department for large stores.

Waterside Number Two

Given the spiraling growth of the electrical load of Manhattan, the demand that Murray had forecast for 1910 was in sight by 1905. While the later turbines produced more than twice the power of the original engines, the practical limit had been reached. An additional station was the only feasible solution. Designed also by Murray,

Waterside Generating Station Number Two opened 11 November 1906. Located on 39th Street to the north of the original station, it built upon the precedents set by Waterside Number One. All of the components; furnaces, boilers, and turbine alternators, were of an improved design. The dimensions were greater than those of the original station, the exterior measured 347 feet 2 inches by 197 feet 9 inches at the longest and widest points. Ninety-six boilers supplied steam to turbine alternators. A total of ten alternators were installed, the first pair rated 7,500 kW, the next six at 8,000 kW. The final pair were rated 14,000 kW. each for a total station capacity of 91,000 kW. In addition to the usual phone lines to the System Operator, the two stations were linked by a pneumatic tube communication system similar to those used in department stores, libraries, postal, and telegraph offices. Advance notice of the increased load for lights caused by the dark skies that accompanied sudden summer thunderstorms was provided by an atmospheric static electricity sensor on the roof of the station.

Reconstruction

Both Waterside stations were expanded subsequently in order to meet the constant increase in the demand for power. With Waterside Number Two complete, the reconstruction of Waterside Number One was initiated in 1911. Three new 20,000 kW turbine alternators were installed in the space occupied by four of the original

IV: GROWTH THROUGH ECONOMY OF SCALE

3,500 kW engine-alternator sets. Each such modification represented a gain of more than four hundred percent, as 14,000 kW of engine capacity was replaced by 60,000 kW of turbines. Subsequent to the completion of the modernization of Station Number One, a similar program was initiated at Station Number Two. By that time, technical advances had produced 30,000 kW turbine alternators that fit within the space occupied by the 7,500 kW machines which had been installed at the time Waterside Number Two was constructed.

Within a few years, the transmission area of the Waterside Station had expanded well beyond Manhattan. 60 Hz capacity was added in 1906 to assist the United Company, the following year that was expanded to carry all the load of United pending construction of a new United station. By 1912, 60 Hz power from Waterside carried the base load of the New York Edison Bronx District, the Bronx territory west of the Bronx River. The 140th Street-Rider Avenue station, constructed for the Bronx District in 1900, was reduced to peak load operation. Some power was transmitted to Westchester to assist in the construction of the Kensico Dam project. In addition the 25 Hz transmission lines were connected to the Metropolitan Street Railway plant on East 96th Street for mutual support. A connection over the Brooklyn Bridge also enabled power exchange with the Edison Company of Brooklyn.

Transmission improvements were frequent. Reactance coils were installed in both Waterside Stations to limit the effects of system disturbances and possible damage caused by sudden current surges. The increased capacity at Waterside was matched by the extension of substations and the construction of new facilities. When new converters were installed at major substations such as those on West 26th and West 39th Streets, the units they replaced were then used at other locations. In some instances, those relocated machines displaced other units to yet a third location. At times, a substantial number of Edison men were involved in the installation, removal, and re-installation of rotary converters in substations around Manhattan. Additional battery capacity was added to the substations.

Railway Power

As noted, the prevailing practice in New York City favored the construction of specialized power stations to supply railways. New York Edison offered in 1909 to supply Waterside power to substations on the proposed Triborough subway. That subway failed to materialize and the routes were divided between existing companies that had constructed power systems of their own. In 1912, Waterside power did begin to supply street railway load when the Third Avenue Railway found it more cost effective to purchase Edison power whenever possible. The railway's Kingsbridge power plant on the Harlem River at 218th Street in northern Manhattan was leased to New York Edison which then supplied 25 Hz power to the substations along the streetcar lines. In that way, the base load was supplied by the Edison Waterside stations while the less efficient railway plant was operated only as needed to support peak load.

The railway benefited financially because Waterside supplied most of the power at lower cost. New York Edison gained a large load for the periods when Waterside operated below capacity. Furthermore, the street railway load was predictable, as it was based on the schedule of the railway. Normal lighting loads varied with the weather, in particular summer thunderstorms produced a sudden demand for light that taxed the best efforts of the power plant and substation crews.

Temporary subway power was provided by Waterside in 1916. Pending completion of new facilities, the Brooklyn Rapid Transit Company erected a temporary substation at Canal Street for new subway lines under Centre Street, Canal Street, and Broadway. Waterside supplied that substation; the deficit in the power available to lower Manhattan was compensated by increased operation of the 1891 Duane Street station, the only old Edison plant that retained direct current generation in place.

Environmental Improvements

Environmental problems did not become a major concern until the 1960s, but coal smoke was a source of urban complaint well before the electrical age. In 1914, water "scrubbers" were added to the smoke stacks at Waterside to remove cinders from the gases. Company legend had it that none other than Thomas E. Murray was crossing First Avenue when the wind blew a cinder into his eye. Annoyed, he hit upon the concept of passing the smoke through a spray of water to cleanse the effluent of heavy particles of soot. While it did not receive much notice, another environmental improvement was initiated in 1914. After a test of nitrogen-filled incandescent lamps in street lights, the arc light met its demise. The harsh, ultraviolet-rich spectra of the arc, with the attendant ozone and hissing noise, ended in 1915 with the substitution of incandescent lamps. Though not as bright as the arc lights they replaced, incandescent street lights in a variety of sizes cast a warm glow over public spaces at night.

V: DIRECT CURRENT CUSTOMER SERVICE PEAK

As the load on the system increased, the economy of scale improved and the per-unit cost of power declined. That reduction in the price encouraged a steady increase in the consumption of power. New York Edison and United Electric Light & Power launched 1917 with a two stage rate reduction plan. Effective 1 January, the rate was reduced from 8 cents to 7.5 cents per kilowatt hour. A similar reduction to 7 cents took effect July 1. Overall, that represented a sixty-five percent reduction in the price of power in comparison to the 20 cents per kilowatt hour rate of 1900. In terms of the urban economy of the time, the 1900 rate was four times the price of a five cent transit fare, cigar, or glass of beer. The rate in the latter half of 1917 was less than one and a half times the price of those same items. Outside Manhattan, the rate was one to three cents higher due to the lower volume of production.

It was assumed that the reductions would continue to stimulate increased consumption and greater economy of scale. That forecast was accurate over time, but short-term increase was constrained by world events. The entry of the United States into World War I produced restrictions on the use of coal and an eventual shortage of manpower. Federal authorities were concerned that the coal demand of war related industries would exceed supply when combined with that of the railroads and steamship lines that were vital to the movement of military traffic.

Electric utilities were ordered to reduce coal consumption. Much of that reduction was met by restrictions on commercial illumination in the evening. Limitations were placed on the hours that large signs and theater marquees could be lighted. The price of coal increased and a "coal" surcharge was imposed on all customers; residential, commercial, industrial, and railway. The manpower shortage was addressed by New York Edison with the training of women as substation operators, the classes held in the school at the West 26th Street complex. Unlike local transit companies that employed women, the company did not assign women to substations before the war ended.

Lost Ground Recovered

The load on the electric system increased rapidly as local economic development accelerated after the war ended. Commercial and residential construction in New York

V: DIRECT CURRENT CUSTOMER SERVICE PEAK

City increased at a rate that exceeded all previous records. Major portions of the outer boroughs were opened to residential development by new rapid transit lines. That produced a population shift which released for commercial development many areas of lower Manhattan that had once been residential. New York Edison conducted a vigorous sales campaign through the distribution of brochures and sales kits. Contractors and builders were offered an illumination guide. A sign planning kit was made available to architects and commercial designers. Residential consumption was encouraged through the promotion of domestic appliances such as refrigerators, washing machines and cooking devices.

The company surveyed 144,000 homes in 1925 to determine the actual use of appliances and lights. That survey indicated that lighting was still the primary residential load. Electric irons were the primary appliance, and toasters were common. Other cooking devices made up a small fraction of the total, while washing and sewing machines were rare. Radios that operated on household power had yet to appear. The survey also explored customer satisfaction related to cost and problems with flickering lights. Advertisements detailed the variety of services available and the company's concern for reliability as well as customer satisfaction, regardless of the size of the account. Many bills to small customers were less than one dollar per month.

The sales force continued to target private plants for replacement with Edison service. Private plants were constructed at the rate of fifty-two per year when Waterside opened. That figure declined in the course of two decades to an average of one per year. That decline resulted from the comparatively low cost of Edison service and the effort of the sales force. Over time, the focus of that sales effort changed. In 1901, the economy of Edison service was promoted as an alternative to the construction of a private plant. Twenty years later, the focus was the substitution of Edison power in place of the private plants, many of which were aged and in need of modernization or replacement. The steady success of the sales effort was obvious in that private plants were canceled or retired at the rate of several dozen annually.

Promotional pamphlets detailed extensively customer satisfaction with the substitution of Edison power for the private plants of offices, hotels, theaters, department stores, and other businesses. Those booklets detailed the reliability and reduced expense of Edison power combined with the convenience of the elimination of smoke, coal dust and ashes. They noted that income could be produced by the space which was released when a private plant was dismantled. Additional space was made available when electric elevators powered by Edison lines replaced the cumbersome plants of hydraulic elevators. Many property owners used the space that was released for storage rooms. Several installed bank vaults, and at least one added a restaurant.

V: DIRECT CURRENT CUSTOMER SERVICE PEAK

System Expansion

Increased power consumption required additional generating capacity. By 1924, the total capacity at Waterside was 362,700 kilowatts, an increase of more than seven hundred percent over that of the original station. At that time, Waterside supplied not only the Edison substations and those of the Third Avenue Railway streetcar system but also those of the Hudson & Manhattan Railroad that is now known as the PATH System. The H&M had leased its power station in Jersey City, N.J. to New York Edison in 1921. Under a contract similar to that signed with the street railway in 1912, Waterside supplied the base load of the substations, the Jersey City station operated only to support peak load. Waterside supplied also the city street lights and the pumping stations of the water mains of the City's high pressure fire hydrant system.

As the demand for power increased beyond the capacity of Waterside, 140,000 kW of 25 Hz power was supplied by the United Company from its new, record-setting, Hell Gate station in the Bronx. United supplied only 60 Hz alternating current to its customers, but installed some 25 Hz capacity to assist Waterside. The Hell Gate capacity was made available by the transfer of the load of the affiliated Queens company (New York & Queens Electric Light & Power Company) from the United system to the Brooklyn Edison system. That allowed two 60 Hz units in Hell Gate to be modified to 25 Hz machines while a pair of 60 Hz units on order were changed to 25 Hz. machines. Overall, the change reduced the amount of 60 Hz capacity available to United customers and those of affiliated companies in Westchester, Queens and the Bronx which were supplied by United. Waterside had a limited 60 Hz capacity which had supplied Edison customers in the Bronx and also the power company in Queens. That was made available to United but the increase in demand throughout the city was forecast to continue.

The only practical solution was additional 25 Hz generating capacity for New York Edison. There was no space for additional construction at Waterside but land was available at a gas plant one mile to the south. The last power station designed by Thomas E. Murray, it was intended to meet the forecast increase in electric load and make possible the retirement of the leased Kingsbridge and Jersey City railway plants and also the last Edison direct current local station still operable. That was the 1891 plant on Duane Street, which was held in reserve. The new plant was one of the largest stations of the time in physical dimensions, and it set records for the size of components and the complex railroad and barge operation that was used to deliver them from the General Electric factory in Schenectady, N.Y. On 23 November 1926, Queen Maria of Rumania threw the ceremonial switch that sent the first power from the new East River station to the substations.

V: DIRECT CURRENT CUSTOMER SERVICE PEAK

Even as the old railway plants were retired, the company acquired in 1927 a pair of stations constructed by the New York Central Railroad in 1906. As in the case of the earlier leases, the railroad found that the twenty-year old stations had been surpassed in efficiency by the Edison stations. Located in Westchester (Glenwood) and the Bronx (Port Morris,) the plants were connected to the railroad substations by a transmission line along the railroad right of way. The Westchester plant was purchased outright, the Bronx station leased as in the case of the other railway plants. Both were connected to the Edison system in Manhattan. That link also enabled mutual support in the event of emergencies.

In time, the 60 and 25 Hz Edison and United systems were linked through frequency changers at the Edison Waterside and East River stations and another at the United Company Hell Gate station. Essentially motor-alternator sets with a common rotor, the machines permitted more precise control of frequency. Early motor-alternator sets used to link the two systems of United and New York Edison were subject to so much variance that New York Edison books issued to contractors noted as late as the 1920s that power in the Bronx was "supplied at a frequency of *approximately* 60 cycles."

Transmission and Distribution Improvement

The capacity of the transmission and distribution system was increased by an extensive program directed by John

Lieb. Like Murray, Lieb rose through the executive levels of the New York Edison Company and became the recognized authority in the United States on the provision of direct-current distribution in urban areas. The capacity of the 25 Hz transmission system was doubled by an extensive program during 1921-22 that increased transmission voltage from 6,600 to 11,400. That program required extensive modification of the entire system of transformers, transmission cables, protective relays and control equipment. Distribution improvements included substation components, cables, meters, and new monitoring and protective systems.

Substation capacity was also increased by the installation of components of greater capacity. Rotary converters installed a decade earlier were relocated as machines of 4,000 kW. and 4,200 kW became the standard for the largest substations of the 1920s. Automatic voltage regulation was used on newer rotary converters, and voltage monitoring was improved in all systems. An indicating system displayed the status of all transmission circuits to the system operator at Waterside. New relays were developed that provided more rapid disconnection of faults on the outgoing feeders. New overload protection was installed in most substations and also in manholes and cable vaults. Circuit breakers were installed in the underground junction boxes where feeder cables were connected together. Intended to prevent "flash over," in which an arc caused by a fault on one cable affected another, the new circuit breakers reduced substantially the number of such incidents. In time more than 3,600 such circuit breakers were installed in the underground junction boxes.

Substations in the 1920s

Land in Manhattan was scarce and even more expensive in the 1920s, but additional property was acquired for substations. Fifteen new substations were constructed and twelve received extensions. Each of the substations had a distinctive appearance, the design often determined by that of adjacent buildings. The vertical format was expanded horizontally as many occupied two or more lots.

West 39th Street

The rapid pace of expansion that characterized the West 26th Street territory in the first decade and a half of the twentieth century was exceeded during the 1920s by the expansion in the area supplied by the West 39th Street substation. It began in November 1923, with increased capacity at the reconstructed 39th Street complex. Rapid development of the garment district and the Times Square theater area required additional facilities almost immediately. The load of new department stores and the millinery businesses between Sixth and Madison Avenues was transferred to a new substation complex on East 32nd Street in October 1923. That substation ultimately comprised three buildings in a 'T' shape, the first two on East 32nd Street and on the west side of Madison Avenue. The last, on East 33rd Street, was added in September of 1926.

In what must have been a record setting expansion of capacity that probably still stands, three new locations were completed in 1925. In January, an "automatic" substation was constructed on the Waterside property at East 41st Street. Similar to a design pioneered in Detroit and another on the New York Central railroad, the automatic operation was controlled by the load and monitored through a telephone line relay from the East 39th Street substation. The new substation assumed some of the load on the East 39th Street substation which freed capacity there to permit East 39th to assume some load of the West 39th Street substation. In June, customers north of 44th Street were transferred to a new substation on West 48th Street that went through the block to 47th Street. Located near the fashionable stores on Fifth Avenue, it was disguised as an appliance showroom with facade windows on the 48th Street side designed by the Gorham Company, a supplier of fine architectural detail. The entire front could be retracted to permit access but that fact could not be detected from outside. A small office handled inquiries and purchases.

In October, a new substation on 36th Street west of Eighth Avenue replaced the substation in Gimbel's department store. Whether the store wanted the space, or whether there was insufficient space for expansion, was

V: DIRECT CURRENT CUSTOMER SERVICE PEAK

not disclosed. Presumably it was a space and capacity issue. Still, the increased demand that resulted from the economic boom was such that additional capacity was required. However, those three were still not sufficient to meet the projected increase in the load imposed by the rapid construction of large buildings in the area.

In May of 1926, a "silent" substation opened in the basement of the Murray Hill Building on Madison Avenue and 41st Street. The operating components were mounted in cork to prevent the transmission of vibration to the building structure. Constructed to meet new demand on the East Side, it assisted the West 39th Street substation by supplying residences, hotels and new stores between Fifth and Madison Avenues. Two years later, an extension of the 36th Street substation was constructed on 35th Street to supply the demand of the garment center.

Overall, in the five years between the fall of 1923 and the summer of 1928, the total connected load and the number and capacity of New York Edison substations in the midtown area multiplied at a rate greater than that anywhere else in the world. Similar expansion of facilities and capacity took place simultaneously in the downtown financial and shipping centers and also in the residential areas on both sides of Central Park. Even so, the successors to the 1888 Edison stations in midtown retained their record capacity. The West 39th Street substation was then the most powerful with 30,800 kW of converter capacity. West 26th Street retained the record for the total number of feeders.

West 26th Street

The West 26th Street substation continued to represent an extreme concentration of substation capacity. Given the concentration of load and the anticipated increase that was forecast as a result of the economic boom of the 1920s, the territory supplied by the West 26th Street station was reduced again in October 1925. The new southern border was 23rd Street; customers on the south side or below were transferred to a new substation on 22nd Street west of Sixth Avenue that went through the block to West 23rd Street. It was operated in later years from the West 26th Street substation.

The significance of the West 26th Street substation was demonstrated on the evening of 5 December 1919. On that occasion, a short circuit in one of the ducts on the 27th Street side flashed over to adjacent cables and involved all forty-four feeders. In the worst disruption to Edison service at that time, all power was cut off for forty-five minutes in the area between 27th and 31st Streets bounded by Seventh Avenue on the west and Madison Avenue on the east. At that time, the station was rated at 21,500 kW supplied by a dozen rotary converters.

Service Variation

Substation loads varied according to the characteristics of the area served. The move of major department stores from the "Ladies' Mile" offset new industrial load in

V: DIRECT CURRENT CUSTOMER SERVICE PEAK

the West Twenties. The new stores in the Thirties were larger and they increased the load on the substations in that area substantially. The load on the substations of the financial district was increased by more than 4,500 kW in early 1929 when brokers installed some 7,500 new high speed stock tickers that consumed 600 watts each. It may be presumed that the load decreased toward the end of that year when the stock market crashed and many brokers left the business.

Substations that supplied newspaper printing plants had to accommodate a unique evening peak load. The newspapers of the city had reached an agreement to start their presses at the same time so that none would gain an unfair advantage in getting their papers on the street. The result was a one hour rush each evening that imposed a load of more than 1,500 kW that continued for an hour with sudden breaks. The longest and heaviest load peaks came on Thursday and Saturday evenings when the editions were thick. Friday papers were heavy with advertising; Sunday papers featured a number of supplements. Substations that supplied bakeries and food processors had to contend with heavy demands during the midnight hours.

Substation Operation

Despite improvement, New York Edison substations were basic installations which were crude by modern standards. Retired employees who operated them in the

1950s and 60s have recounted their experience with the primitive equipment. The converters had to rotate in synchronism with the frequency of the applied current. Some were started from the incoming 25 Hz lines; others were started from available direct current in the substation and then synchronized with the incoming power. Failure to synchronize properly would cause the machine to "hunt" or change speed rhythmically. As for the storage batteries, old photographs show rows of bins the size of modern refuse dumpsters that were filled with lead plates and sulfuric acid. The presence of that much acid in an enclosed space must have produced an interesting flavor to the air of substations which was filled with the tang of ozone from the intense arcing of the converters when they were started.

The outgoing feeders were connected between substations to assure the continuity of service in the event of substation failure. In the event of a failure of the supply from the power stations, the substations could not be started individually upon restoration of power. To do so would have produced heavy overloads on the first to restart as a result of the interconnection of the feeders between the substations. Instead, the substation operators were instructed to start and synchronize their converters without connecting the load. When all the substations were ready to assume load, the System Operator would sound a bell to signal all substation operators to connect the load simultaneously.

V: DIRECT CURRENT CUSTOMER SERVICE PEAK

The only known instance in which that procedure was used occurred in November 1965 when a regional failure of the 60 Hz system spread to the 25 Hz lines through frequency changers that linked the two systems. All the substations shut down; restoration required the established procedure. Under such circumstances, there still must have been momentary surges and arcing to an extent never seen previously.

Tradition Ruled Manhattan

In 1928, it appeared that New York Edison intended to continue indefinitely the extension of direct-current distribution. An extensive feature in the Edison Weekly of 1927 detailed the reliability of direct-current systems with battery reserve. Plans were prepared for at least two new substations and one extension to an existing location. In an address in June 1928, New York Edison Vice-President John W. Lieb stated that direct current distribution systems were preferable in areas such as the midtown and the financial district of Manhattan. He described the heavy load imposed by factories, printing plants, restaurants, theaters, and large office buildings with thirty or more elevators. Regardless of appearances, major change lay ahead as the capital and operating expense of new substations had become a concern.

VI: ALTERNATING CURRENT INCURSION

As the expansion of direct current distribution dominated the midtown area during the first two and a half decades of the 20th century, the United Electric Light & Power Company continued the relentless determination of Westinghouse to establish alternating current as the standard. The original United 133 Hz system was not efficient and did not earn a reasonable return on investment. Prior to the mid1890s, the primary source of United revenue was that of the arc light business of its predecessor company; a business which expanded with the acquisition of the Brush Company in 1892. The first United incandescent lighting system that earned income sufficient to cover expenses was installed in the mid 1890s, the frequency having been changed to the standard 60 Hz about 1893. Featured in *Electrical World*, a power station on East 29th Street at the East River was reequipped with 60 Hz alternators salvaged from the Chicago Columbian Exposition of 1893. Much of the early innovation at United was directed by W.E. McCoy, an engineer who had collaborated with Westinghouse chief engineer Benjamin G. Lamme on the development of much of the pioneer equipment.

Rebuilt from an arc light station, the East 29th Street station supplied two-phase power, the standard for 60 Hz alternating-current distribution at the time. There were exceptions to that practice, the Municipal Electric Light Company of Brooklyn installed in 1897, a 60 Hz three-phase system in the station at Rodney and Ainslie Streets in Williamsburg. It supplied a number of induction motors in the factories of the Greenpoint area and was probably the first three-phase 60 Hz distribution in New York City. Power was said to have been transmitted at 2,080 volts single-phase, 2,300 volts three-phase. However, the Brooklyn installation supplied primarily industrial load, whereas United intended to supply a customer mix that included substantial residential and commercial single-phase loads.

In 1896, the company supplied the first two phase "power service" to a 15 hp pump with a two phase wound rotor motor; the location was not specified in the report. Early records show conflicting details, but it appears that by March 1898 the United Company was the first to operate steam-driven alternating-current generators in parallel, a task vital to large scale operation but accomplished only with hydroelectric plants prior to that time. United extended the two-phase system north of 59th Street in 1899 and on 26 December of that year connected to a customer on Dyckman Street at the far northern end of Manhattan Island. The extension of United power

VI: ALTERNATING CURRENT INCURSION

to that area, which was almost rural in character at the time, displayed the determination of United to provide a truly comprehensive system that would exceed the limitations of the direct current distribution system of the Edison company. United lines then extended the length and breadth of Manhattan Island. Transmission/distribution voltage was not specified in the report.

Below 59th Street, new lines were overlaid on the previous system to provide sufficient capacity for new industrial customers on the lower West Side. The generating capacity at East 29th Street apparently reached 7,200 kW by 1905. It appears that power was transmitted from the 29th Street station at 7,500 volts, 60 Hz. to a pair of transformer substations, one a former arc light station on Elizabeth Street in lower Manhattan, the other on 146th Street in northern Manhattan. A direct 7,500 volt line connected to a private transformer station in the National Biscuit Company plant at Tenth Avenue and 15th Street, the largest industrial customer in the city.

The United Company system distributed power throughout the area of Manhattan south of 59th Street with underground cables, the original overhead system having been placed underground in 1889. Early records are not clear; transmission/distribution voltages of both 2,300 and 3,000 are mentioned as is 4,160 in at least one reference. By 1906, 3,000 volt primary distribution was established. Those lines extended to distribution "points" from which smaller 3,000 volt lines of less current capacity connected the transformers that supplied the customers. Regulations of the Central Telegraph & Electrical Subway Co. that owned and maintained the street ducts forbade the installations of transformers therein. Therefore, transformers were located in street vaults or building basements to supply customers via a four wire circuit with either single phase "lighting service" at 110 volts for residential and light commercial needs, or two-phase "power service" at 191 volts for heavy commercial and industrial loads. The two-phase system was simplified in the early years of the century by interconnection of the phases to permit the use of only three wires. Single-phase "lighting service" was then increased to 115 volts, while two-phase "power service" was supplied at 199 volts. It appears that the United distribution voltage was standardized at 3,000 volts by that time.

By contrast, most 60 Hz alternating-current distribution in United States retained the 2,300 volt standard in developed areas, where transformers were spaced at close intervals. One of the few exceptions was the Edison Electric Illuminating Company of Brooklyn which in 1897 pioneered the use of 11,000 volt transmission/distribution because of the distances covered. The station, said to be the first to use the new standard in steam driven alternators, was located at 66th Street on the Bay Ridge waterfront. Often called the Union station because it was intended to supply power to any utility company in Brooklyn, it supplied two separate transmission systems. One was a three-phase, 25 Hz, 6,600 volt system connected to direct-current substations in the developed

VI: ALTERNATING CURRENT INCURSION

areas of Brooklyn. The second was a two-phase, 62.5 Hz, 11,000 volt line to the alternating-current distribution transformers in the sparsely developed sections of southern Brooklyn. There may have been an intermediary 2,300 or 4,160 volt secondary distribution system to the transformers that supplied the customers. Again, records are not clear. The odd 62.5 Hz frequency was a function of the provision of both 25 and 60 Hertz power in a common alternator frame. A separate system on Staten Island used 5,000 volt transmission the length of the island from a 60 Hz station which opened at Livingston in 1897.

Like New York Edison, the United Company conducted vigorous promotion efforts. In 1905, that campaign had two distinct objectives: residential customers in northern Manhattan and businesses throughout the island. It aimed at residential customers through billboards and banner signs at major street intersections. The business of small merchants was sought by the provision of free electric signs. Those signs were expensive, with multiple sockets and ornate scroll work detail. The company declared that the investment paid for itself within a year as most of the merchants would not have installed an electric sign without that incentive. Newspaper advertisements were backed by a promotional booklet *Lights & Shadow*s that was distributed to customers, architects, contractors and realtors. The promotion succeeded as United obtained the business of industrial customers on the Lower West Side and world famous lighted signs in Times Square.

So rapid was the development of northern Manhattan after the extension of the new subway line to that area in 1906 that a branch office and showroom was opened at Hamilton Place near 143rd Street and Broadway in September, 1910. With a street frontage along most of the block, the showroom offered a variety of electrical devices for the customer to try before buying. Sales of electric irons were promoted by a program in which they were loaned to customers for a thirty day period at the end of which the customer had the option of buying or returning the device. It was estimated that sales of electric irons exceeded one hundred units per month.

The success of United in the development of industrial load probably rested in part on the extensive promotion of improved alternating current motors by Westinghouse and other manufacturers. Unlike many utility companies that supplied alternating current to customers, United did not put restrictions on types and designs of motors for elevators or other machines but required that only a few basic standards be met. United captured the business of bakeries, stone masons, leather goods factories, and the manufacturers of welded tools. The success of United in the animated sign market was largely because of the fact that the complex automatic switches needed to produce the animated effects were subject to less wear from arcing when operated on alternating current. That allowed the use of smaller, less expensive switch contacts that lasted longer with less maintenance.

VI: ALTERNATING CURRENT INCURSION

United Incursion

Ironically, United's largest industrial customers were located in the territory supplied by Edison direct current from the West 26th Street substation and later additions; a fact which illustrates well the sheer load density of the area. The heaviest United load was that of one hundred seventy-four Westinghouse two-phase motors in the National Biscuit Company plant that occupied the entire block from Fifteenth to Sixteenth Street between Ninth and Tenth Avenues. The Kay-Scherer Surgical Instrument Company on Ninth Avenue at 27th Street operated seven General Electric two-phase industrial motors and also alternating current elevator motors. The largest alternating current sign in that territory was "U-Need-A Biscuit" at 16th Street and Eleventh Avenue.

United supplied many large signs located within the territory of the Edison direct current substation on West 39th Street and later additions. A large sign promoted sparkling water at Broadway and 35th Street. Two blocks away at Broadway and 37th Street was the world famous Victor "Talking Machine" sign with moving letters. In the Times Square area, the largest signs were the "El Bari Gin Rickey" sign at Broadway and 42nd Street, and Levey's "waving feather" a block to the north. The largest and most controversial was a fifty foot-wide animated Heatherbloom Petticoat "Silk's Only Rival" sign. 1,837 lamps depicted a girl in a storm, her skirts battered by wind and rain. The statistics were staggering.

Mounted upon a fifty-five foot wide frame were the outlines of a seventy foot tall skirt with a fifty foot wide bloom; the shoes measured twelve feet long with three foot wide buckles. Overall, the dimensions were more akin to those of the balloons of a modern Macy's Thanksgiving Day parade than an advertising sign. The sign was deemed either fascinating or gaudy, the opinion dependent on the eye of the beholder.

Power Generation

Additional generating capacity was required to meet the new demand but expansion was constrained by the size of the East 29th Street station. As United became a victim of its own marketing success, assistance was provided by the installation of motor-alternator sets (25 Hz motors connected to 60 Hz alternators) at the New York Edison Waterside station to supply an additional 4,800 kW of 60 Hz power to United. Waterside became the sole supply of power to the United Company on 18 November, 1907 when the East 29th Street United power station was closed after the property was taken by the City of New York for an extension of Bellevue Hospital.

At that point, a pair of 7,500 kW 60 Hz alternators was installed at Waterside # 2 to replace the East 29th Street station while the motor-alternator sets were relocated to both the United Elizabeth Street and the 146th Street transformer substations and connected to Edison 25 Hz lines. The 60 Hz United alternators at Waterside

VI: ALTERNATING CURRENT INCURSION

were mounted on the same shaft as were previous NY Edison 25 Hz alternators. Whenever the 25 Hz machines were operated as generators synchronized with the other 25 Hz units of the Edison system, the design of the United machines was such that they produced 62.5 Hz Such change was acceptable in an era prior to interconnection of systems and the use of electric clocks regulated by line frequency.

The operation of the United system at that time provides an interesting glimpse of the control and management techniques used by the pioneers of alternating-current distribution in the early years of the 20th century. Between the hours of 11 pm and 4 am, the frequency of the United system was 62.5 Hz as the 25 Hz units were also generating power that was synchronized with the other 25 Hz alternators. During the peak load hours between 4 pm and 11 pm, the pair of 25 Hz NY Edison system alternators on the same shafts as the United machines were shut down and the turbine speed was adjusted to permit the United alternators to generate 60 Hz power that could be synchronized with the 60 Hz output of the motor-alternators in the United transformer substations. Those motor-alternators were thus operated only to meet peak demand on the United system.

Records indicate that the motor-alternator sets were not used extensively, most probably a result of the availability of an additional 5,000 kW 60 Hz alternator in Waterside # 1. That unit apparently predated the United connection, most likely having been installed to supply NY Edison customers in the Bronx.

The placement of the two new United 60 Hz Waterside alternators on a common shaft with NY Edison 25 Hz machines also allowed operation of the units as motor-alternators without steam during off-peak hours when spare 25 Hz capacity was available. In that instance, power from the station's largest and most efficient 25 Hz machines ran the 25 Hz NY Edison alternator as a motor to turn the United 60 Hz alternator. That effectively turned the pair of alternators into a 25 Hz/60 Hz motor-alternator set that supplied 62.5 Hz power without steam from the boilers. Thus coal was saved and the "heat rate" (the standard measurement of power station efficiency) was optimized.

Alternating Current Transformer (Sub) Stations

Data on the distribution system prior to 1906 is minimal, but by that point, the United distribution system operated the aforementioned transformer substations with attendants to monitor voltage regulation and power factor compensation. Simpler than the Edison substations, the United transformer stations were equipped with controls, switches, induction regulators to control voltage, and synchronous condensers to regulate power factor. The Elizabeth Street facility was rated at 8,250 kW in transformer capacity, while the 146th Street location was

VI: ALTERNATING CURRENT INCURSION

rated only 1,500 kW in transformer capacity because the northern load was still light by comparison. Those numbers changed substantially with the installation of the motor-alternator sets in 1907. Those motor-alternators, installed initially at Waterside to supply 60 Hz power to United when the projected load exceeded that capacity of United's East 29th Street station, increased substantially the power available from the United system.

The United load increased twenty percent in the six month period after the closure of the East 29th Street plant. That increase was caused in part by residential development in northern Manhattan following the completion of the first subway line. The load in that area continued to grow exponentially as residential and small commercial development occupied space which had been empty lots and fields just a few years before. An article appeared in the *GE Review* (a General Electric company magazine) in 1908 that detailed the United system as of 31 December, 1907. At that time the United Elizabeth Street transformer substation was rated at a total of 10,250 kW. Of that system total, 8,250 kW was supplied from the 60 Hz transformers that were powered by the 7,500 volt feeders that were installed to connect to the 60 Hz United alternators at Waterside when the United East 29th Street station was closed. The additional 2,000 kW was supplied by a pair of 1,000 kW motor-alternators which had been relocated there from Waterside and were supplied from the local 6,600 volt 25 Hz lines of the Edison Company.

The 146th Street transformer substation capacity increased substantially to a total 4,800 kW of which 2,000 kW was supplied by the aforementioned 60 Hz 7,500 volt feeders that were connected to the United 60 Hz alternators at Waterside when the United East 29th Street station was closed. An additional 2,800 kW was available from four motor-alternators relocated to 146th Street from Waterside; two were rated at 1,000 kW, the other pair rated 400 kW. The source of 25 Hz power for those motor-alternators was not stated, presumably feeders were extended from the northernmost Edison 25 Hz. lines.

United Electric Light & Power Co. transformer substation control room West 146th Street Substation

VI: ALTERNATING CURRENT INCURSION

During the off-peak period, the idle motor-alternator sets in the transformer substations were operated from the 60 Hz side as synchronous condensers to supply reactive power into the United lines. Reactive power was a concern of particular importance when load was minimal and the reactive power consumed by transformers became a larger percentage of the total power and thus reduced power factor. That practice was common at the 146th Street transformer substation in particular as it was seven miles from Waterside, and any improvement of the power factor on the distribution system in the northern area yielded significant savings by elimination of a "wattless" power component on the transmission lines. While arcane by modern standards, it was a vital step in the development of alternating current distribution practices.

The *GE Review* stated that the system supplied a total of four hundred and twenty elevators powered by two-phase alternating current motors, an unusual application as direct-current motors were the standard for elevators at that time. It was said that the customers were pleased with the elevator motors because the controls and maintenance were simple. The primary 3,000 volt distribution system from the transformer substations at that time used three lines, so arranged that the "outer"pair was energized at 3,000 volts while the potential between either of those and the third was 2,100 volts. At the distribution points the three-wire scheme was continued to customer transformers with two-phase motor loads; while only the 3,000 volt pair was continued to the customer transformers for "lighting" loads. That configuration simplified the system, and facilitated voltage regulation. At that point the company had one hundred and fifty miles of street lines with three hundred and fifty miles of duct work under lease and more than nine hundred and fifty miles of 3,000 volt distribution lines that supplied 1,356 distribution transformers for lighting customers, and an additional 420 distribution transformers for power (motor) customers. Total system capacity was 12,000 kW. Customer meters totaled some 14,000.

Although less efficient than later practice, the system provided practical alternating-current distribution in an urban area. Moreover, the placement of transformers in underground ducts was prohibited, while fire codes forbade installation of oil-filled transformers inside buildings. United installed clusters of air-cooled transformers in buildings wherever property owners permitted. Those units often supplied not only that location, but nearby structures with cables that "looped" back into the street ducts to reach the adjacent customers. Nevertheless, ventilation, transformer size, available space, and access pathways imposed restrictions on the total capacity that could be installed in any one structure. United Company distribution engineers were thus forced to devise innovative solutions to meet the needs of the customers.

VI: ALTERNATING CURRENT INCURSION

Three years later, *Electrical World* reviewed more extensively the United system in an article in its May 1911 edition. It began by noting that, had United been located in another city, it would have been considered among the largest utility companies in the United States. However, the Edison name was so well known in New York City that the significance of the developments of the United Company was largely overlooked. By that time, the total number of alternating-current elevators exceeded five hundred and fifty, many of them in the new large apartment buildings constructed along the subway route. Some of the apartments contained as many as eight, or even twelve, rooms and constituted a substantial single-phase "lighting" load. Overall, the company was supplying a total of 1,100 large apartment buildings.

Industrial load had continued to increase to comprise fifteen percent of the total, and continued to be led by the National Biscuit plant, which had increased the number of motors to two hundred in sizes from one third to 150 horsepower. The internal transformer substation was rated at 1,200kW, supplied via a direct 7,500 volt transmission line from Waterside. It was the largest industrial customer on any utility system in Manhattan. At that time, United had 25,000 customer accounts, a phenomenal increase brought on by the residential and commercial development of northern Manhattan. The corporate office was located in midtown at 1170

Broadway at 28th Street. United also supplied substantial power to the city for the illumination of streets, buildings and various motors. Total revenue reached two million dollars annually. Supervision of the United system was directed by the System Operator at Waterside which remained the sole source of power for United until 1913

The Sherman Creek Station

After an unusually long delay, which was most probably caused by the economic situation that prevailed nationally after the financial "Panic of 1907," United initiated construction of a new station that had been planned as early as 1905. The dependence on New York Edison Waterside power station ended with the opening of United's Sherman Creek station in October 1913. Located in northern Manhattan and named for an inlet off the Harlem River, the station became a leader in the development of 60 Hz power plant operating protocols. The plant improved operating techniques to an extent well beyond those established at Waterside, for Murray had continued to build on previous advancements. His design advances enabled a substantial increase in the capacity of boilers compared to those installed at Waterside only a few years previous.

The power station electrical system featured improved control equipment and reactance coils to limit potential damage to generators and transformers in the event of transmission short circuits or surges. It was said

VI: ALTERNATING CURRENT INCURSION

that it was the first to use a water spray system to remove ash from the smoke (the scrubbers at Waterside were added one year later in 1914), the first to use pulverized coal, and also the first to use an idle turbine-alternator set as a large synchronous condenser for voltage support through improved system power factor. Sherman Creek also supplied power to affiliated companies in the Bronx and Westchester. At that time, the 7,500 volt transmission system was connected to Sherman Creek and the 3,000 volt distribution system remained in place.

The number of customers had increased to more than 34,000 by 1914, a phenomenal increase of some thirty-six percent in less than three years. As development in northern Manhattan continued, United constructed a new transformer substation on 187th Street to supply the area. The new structure contained not only the electrical equipment but also a library, classrooms, and various field offices. Supervision of the United Company system remained under the direction of the Waterside office.

National Trend

The second decade of the new century was marked by a national trend to install alternating-current distribution wherever it was practical. Years of effort to improve 60 Hz distribution had made it practical in urban areas of moderate load concentration. A number of utility experts stated that universal, standardized 60 Hz alternating current transmission and distribution was on the horizon. That trend accelerated as the number of installations increased. Once a large area was supplied by 60 Hz lines, the need to maintain 25 Hz generation and transmission systems plus the substations required for direct-current distribution posed an expensive complication. As urban load increased after World War I, the extension of direct current systems was expensive and complicated. Extension of 60 Hz alternating current systems involved only the installation of new transformers and lines. It also offered an advantage in urban areas with extensive piping for water, gas, and other utilities as it did not produce the electrolytic corrosion damage that was caused when stray direct current passed through those pipes.

The U.S. Commerce Department had encouraged the precise regulation of frequency as a means to ensure good power factor, and thus efficient operation which would conserve coal during the first World War. That effort also acted to encourage the interconnection of systems to share generating capacity and provide mutual support. Shared capacity increased the available reserve which could be pooled to meet peak demand. A portion of that reserve could be "spinning;" that is the alternators were synchronized but not loaded. With sufficient reserve to protect against failure of the alternating current system, the advantage once offered by the substation battery reserve of direct current distribution systems was obviated.

VI: ALTERNATING CURRENT INCURSION

Nationally, the interconnection of 60 Hz generation and transmission systems developed rapidly. The pace of interconnection accelerated throughout the economic boom of the 1920s. John W. Lieb detailed the trend in an address to utility experts in June of 1928. He stated that the 60 Hz power systems throughout the eastern third of the United States were synchronized and interconnected to form a web that extended from the Great Lakes to the Gulf Coast between the Atlantic seaboard and the Midwest. It comprised seventy percent of the generating capacity of the United States and enabled the exchange of power between localities and regions to produce an economical system that was beneficial to all involved. The most efficient stations carried the base load; older plants of lesser efficiency were called upon only when demand required.

The major manufacturers of electrical components favored the universal use of 60 Hz alternating current because it offered vast savings through the standardization of product lines. In 1915, Westinghouse introduced an improved distribution transformer suitable for installation in basements and street vaults. It was a sealed unit of high efficiency that was adequate for the load density found in Manhattan. Similar products were introduced subsequently by other manufacturers. United was often the focal point of Westinghouse efforts to develop new methods and improve operation for it was a unique urban test bed that employed only alternating current distribution. By comparison, the power companies in most cities operated both alternating and direct current distribution systems, the type dependent on the concentration of load in a particular area.

Induction motors had undergone similar improvement to the extent that low power factor could no longer be cited by proponents of direct-current distribution. In point of fact, all types of alternating current motors had been refined substantially in the early 1900s. Steinmetz led the research at General Electric and innovative designs were also developed by companies such as Allis-Chalmers, Fairbanks-Morse, Wagner and Reliance. In point of fact, those smaller companies pioneered several significant improvements in alternating current motor design.

By 1920, alternating current motors had reached the point that they equaled direct current machines in most industrial and commercial applications. Direct-current motors continued to dominate only the railway and elevator business. Technical evolution occurred in smaller applications as well. By the mid-twenties, small alternating-current motors had been developed which made possible compact appliances and office machines. Some of those incorporated concepts developed by Steinmetz as a result of his research on the Teaser circuit of his early Monocyclic system. Electric clocks with motors regulated by the frequency of the line kept time with less than one second error in five months.

VI: ALTERNATING CURRENT INCURSION

Regional Interconnection

Interconnection of 60 Hz systems in New York City made possible substantial economies through the elimination of small power stations. In the 1920s, the 60 Hz capacity of the United, New York Edison, and Brooklyn Edison Companies was pooled to supply customers of "affiliated" companies in the Bronx, Queens and Westchester County. Small stations of low efficiency in Queens and the Bronx were closed. By 1929, the pooled capacity was such that a pair of large stations was deferred until 1940 at the earliest. The total installed generating capacity of the three companies was 1,757,000 kW, half of the total of New York State, which included a new large hydroelectric station at Niagara Falls. It was said to be sufficient to supply any of the forty-four mainland states with the exceptions of New York, Pennsylvania, Illinois, or California.

The United Company was a major figure in that total, having completed in record time a new power station in the southeast Bronx. Named Hell Gate for the treacherous waters where the East River meets Long Island Sound, it was another advanced Murray design. One of the largest power stations anywhere, it featured thin water-cooled steel furnace walls instead of the thick fire brick used previously. That change increased substantially the useful furnace space. It was also the only utility station that could receive coal by either rail or barge. Intended to supply power to United customers as well as customers of affiliated companies in Westchester, the Bronx,, and Queens, the first power was generated in November 1921, less than two years after construction commenced.

In 1929, electric power for the utility customers of New York City was supplied by six major and two minor power stations owned and operated by the three electric companies. Half of each category supplied 60 Hz power to customers, the balance was 25 Hz for conversion to direct current. Railway load was supplied primarily from eight 25 Hz plants, most of which were linked by tie lines to the Edison generating stations. Much of the increased capacity was attributable to advances in turbines and alternators. The largest units installed in 1920 were rated at 30,000 to 35,000 kW. Five years later, the standard was 50,000 to 80,000 kW. By the end of the decade, machines rated 160,000 kW were under construction. As a result, the United Company Hell Gate station of 1921, though designed for a maximum 280,000 kW was rated 605,000 kW upon completion in 1928. Brooklyn Edison's 1924 Hudson Avenue station, intended for 400,000 kW. ended with a rating of 770,000 kW. in 1932. It then took the title of "world's most powerful steam operated generating station" from Hell Gate.

As United continued to expand after World War I, it improved the transmission system and doubled capacity with an increase in transmission voltage

VI: ALTERNATING CURRENT INCURSION

from 7,500 to 13,200 volts. Three-phase transmission and distribution systems were installed by United in 1923 to increase capacity an additional fifty percent. The new system balanced single phase residential and light commercial loads evenly on all three phases. That substitution was a national trend. Development of new mathematical equations by Charles Fortescu at Westinghouse and Edith Clark at General Electric made possible system design that assured equal division of load across all the individual phases and thus ended the preference for two-phase 60 Hz distribution. An *Electrical World* survey in 1925 disclosed that the use of two-phase motors nationally had declined forty percent over a five year period. The poll of 3,130 companies in fifteen industrial categories found that eighteen percent of the polyphase A.C motors in use by those firms in 1920 were two-phase. In 1925, only ten percent of those motors were two-phase.

The affiliated New York & Queens Electric Light & Power Company was the first in the city to change 60 Hz distribution to three-phase in 1919, the move required to accommodate the load of new industrial plants in Long Island City, Queens. Brooklyn Edison initiated the change in 1923 to accommodate spiraling residential and commercial demand. It appears that the Bronx district of New York Edison adopted three-phase lines when underground lines replaced overhead in the years after the territory was acquired in 1899.

United Midtown Presence Expanded

As noted frequently in *Electrical World*, United led the development and promotion of alternating-current distribution, and it continued to challenge the dominance of the New York Edison direct-current system. In 1914, the magazine again declared that "were it located in some other borough away from the hypnotic spell of the well-advertised word "Edison," United would be recognized as one of the great systems." It noted that United produced more power than the entire state of Rhode Island as it continued to increase substantially the number and variety of customers in midtown Manhattan.

United established a new midtown presence with a transformer substation located at 354 West 45th Street. Placed in operation in October 1915, it was reviewed in detail by *Electrical World* in 1916 in an article entitled "A Compact Alternating Current City Substation." It supplied luxury apartment buildings, theaters, movie "palaces," the new Abercrombie & Fitch department store, the "Iceland" skating rink, and parking garages with automobile elevators. United continued to lead in the provision of power to the major animated advertising signs. The Wrigley's Gum sign was considered the world's greatest with 15,000 lamps and changing colors. A full block in length; it stood 34' high and spelled out three product lines alternately: "Spearmint," "Doublemint," and "Juicy Fruit." The Fisk Tire sign enjoyed equal fame

VI: ALTERNATING CURRENT INCURSION

as it depicted the company logo of a sleepy boy holding a flickering candle with the slogan "Time to Re-Tire."

The 45th Street transformer station also eliminated a facility on 24th Street near Tenth Avenue which appears to have been a transformer substation in a former arc light company station. It may have been a part time or temporary facility as it does not appear in most system descriptions and disappeared entirely from company lists in the early 1920s. An additional transformer substation was constructed on West 97th Street in 1922 to meet demand from new residential construction on the upper West Side. The 97th Street facility was equipped as a three- phase installation as part of the planned transition of older properties from two-phase to three-phase systems. The Elizabeth Street transformer substation was expanded with a new transformer building on the north side of the original structure which had been rebuilt from an arc light plant. The original structure was subsequently replaced by a new and larger substation of the standard design.

Theaters throughout Manhattan were supplied by United. It was noted in advertising that the Loew's Theater at 86th Street and Third Avenue was the source of a complimentary letter from company founder Marcus Loew to the directors of the United Company in regard to his satisfaction with the quality of service and price. United also continued to attract new business of lower west side industries; two of the largest were on West

23rd Street, the Westinghouse Lamp Company factory and the Morgan Crucible Company. The latter made the carbon "brushes" that were small but absolutely vital components of alternators, generators, rotary converters and all motors except those of the induction type.

Other customers included meat packers, refrigerated warehouses, laundries, and dry cleaning plants. Short-term customers that required substantial power for a limited number of months were the contractors engaged in the massive subway construction program known as the "Dual Contracts" that was undertaken in 1913 and largely complete by 1920 . The downtown financial district had been the site of Edison's first station and the first area of the city to be extensively equipped with Edison power lines. The numerous large office buildings of the area represented a dense concentration of load supplied by several Edison direct current substations. That did not deter United from seeking new customers in the area. One of the largest was the Waterman Building known as the "Pen Corner" at Broadway and Dey Street, the headquarters of the nationally recognized manufacturer of Waterman Pens.

The inherent efficiency of alternating current distribution was demonstrated by the simplicity of the system. United supplied more than fifteen percent of the utility customers in Manhattan but required only five transformer substations over the entire length of the island. By contrast, the Edison Company supplied more than four times as much power but required eight times

VI: ALTERNATING CURRENT INCURSION

as many substations to provide power to its customers, who were all located south of 135th Street. Furthermore, the majority of the Edison substations included extensions. Thus the total number of Edison structures was more than seventy-five, while there were only six United Company buildings.

United Promotion and Advertising

While technical aspects were being resolved in laboratories and test installations, United marketed the advantages of alternating current intensively with efforts that rivaled those of New York Edison. It too had a cartoon character, a boy dressed in the style of Aladdin but without a lamp. Instead he reached for a wall switch with a Genie holding a motor or other appliance with the slogan "Like the Genie of Aladdin, Instantly Responds to the Commanding Touch." The most important and informative advertisement was *United Service*. More impressive than the usual brochures produced by most utility companies, it was a monthly magazine with color covers and inserts. Selected Westinghouse appliances were offered at a discount each month, and United alternating-current service to industries, businesses, and homes throughout Manhattan was detailed with photographs and stories. Other United advertisements stated clearly the intentions of the company to establish a system that would compete on an equal basis with the direct-current distribution of N Y Edison. That challenge was declared with banner headlines that stated that United was "Aladdin's Modern Genie" on one line, followed by a headline below in larger typeface that read: "Supplying Electricity for Manhattan's Every Need." An additional office and store was opened at 89th Street and Broadway in May of 1916 to serve the customers on the upper west side.

United "Electric Shops" retail appliance store of the United Electric Light & Power Co. at 138 Hamilton Place

Newspaper advertising focused on the convenience and safety of electric lights and appliances. Displays at the annual electric show hammered home the theme of the efficiency of alternating current equipment with "United Alternating Current Service" spelled out in lights. Large billboards offered the image of a young

VI: ALTERNATING CURRENT INCURSION

woman with various alternating-current appliances and the slogan that they would "Save Work, Preserve Youth." The Sherman Creek station was marked by a large electric sign that proclaimed the name of the company, while bright electric signs adorned the facades of the company offices and shops. On the roof of the uptown office a ten-foot outdoor clock kept perfect time, regulated by the alternators at the power station. Those in turn were regulated by a master clock set by telegraph signal from the United States Naval Observatory. The electric show in the fall of 1917 featured new electric clocks developed by the Warren Self Winding Clock Company.

Resolution of the Final Questions

After three decades of debate, the proponents of universal alternating current distribution were vindicated by two developments of the United Company. The first was an "automatic" distribution network in which voltage regulation and power factor correction required no intervention by an operator. The second was the elimination of transformer stations by connection of power station transmission lines directly to the automatic distribution networks. The network concept was vital to the substitution of alternating current distribution in place of the direct current systems which had dominated much of Manhattan for more than three decades.

Prior to the development of the United automatic network, alternating-current distribution required not only the attended transformer stations, but also duplication of lines and manhole switches to maintain supply in the event of a cable or component failure. A concept that dated to the lines of the 1890s, the distribution system consisted of 3,000 volt cables that radiated from the transformer substations (originally from the power station) to a total of seventy "area distribution points" where smaller cables branched to the transformers that supplied the customers. A total of two hundred and fifty sets of "manhole switches" enabled a fault to be bypassed, but it sometimes required half an hour or more to locate the problem and dispatch a crew to the proper manhole to operate the switches. Expensive in first cost, that cable arrangement did not compensate low power factor. Reactive power was compensated primarily in the transformer substations that were often a distance from the load with the result that distribution cable capacity was reduced by the "wattless" component carried by the lines. It was far too complex to replace the direct current feeder system.

A series of articles in the Westinghouse publication *Electrical Journal* in 1925 detailed the extensive efforts by utilities and component manufacturers to perfect an efficient alternating-current distribution network. The first network, an overhead system, was installed in Peoria, Illinois in 1915. The Commonwealth Edison Company of Chicago installed a similar scheme for underground distribution lines in 1919. Subsequently, virtually every utility that supplied urban service

VI: ALTERNATING CURRENT INCURSION

designed a network that would meet the needs of a particular locale. However, all of them were deficient in one or more aspects that involved reliability (continuity,) efficiency, cost of installation, simplicity of maintenance, automatic "throw-over" (ability to switch out faulted cables,) and the capacity to permit faults to "burn" clear.

The United Company pursued extensive research and development efforts in concert with the research teams at Westinghouse until a concept was developed that would address all the potential problems. On 12 April 1922, United activated an "automatic network" that supplied power to a few blocks on the upper west side of Manhattan. It was a radical undertaking and supervision was maintained through the West 97th Street transformer substation of the United Company. The scheme employed four sets of transformers so arranged that the load was optimized with respect to the transformers. Good power factor was provided by the selection of transformer impedance (originally 8.7% later 10%) so as to force equal division of load between individual transformers. Automatic switches adjusted voltage and a series of relays coordinated operation. Rated at 300kW, it supplied a residential area with fifteen elevators with the customer voltage variation limited to 3% or less.

By 1923, the engineers of United and also those at Westinghouse were satisfied that their development could carry any urban load. The concept, known as the

United network, or more commonly, the New York network, became an industry standard. Vital to the provision of power in urban areas, that network development has been accorded little notice, for it is not obvious to the average person, while the principles involved are understood only by those with an understanding of alternating current distribution concepts. At that time, the United Company was supplying 82,150 customers of which 56,277 were in northern Manhattan. Still the remaining 25,873 customers below 135th Street represented a significant incursion into New York Edison territory.

Three years later, the second innovation connected the 13,200 volt transmission line from the power station directly to the transformers of the automatic distribution networks. Those transformers in underground vaults and building basements were the only link; no additional equipment or space was required. In time, the five manned transformer stations were retired or used only for automatic equipment. The continuity and reliability of alternating-current distribution surpassed that of direct current while the capital and operating expense of the new United alternating-current network distribution system was but a minute fraction of that of the Edison direct-current distribution system with conversion substations. When alternating-current distribution proved superior in the dense concentration of load of midtown Manhattan, it proved practical anywhere.

VI: ALTERNATING CURRENT INCURSION

Transformer for Automatic Distribution Network developed by the United Electric Light & Power Company

The United network provided power at 120 volts single phase for residential and light commercial loads, and at 208 volts, three phase for heavy commercial and industrial demand. That voltage was selected because lights performed best at 120 volts or greater, while motors ran more efficiently with three phase voltage below 215. The lower motor voltage increased marginally the active current as compared to the reactive current and thus improved the power factor of the motors. Distribution efficiency in the large buildings of midtown was improved considerably by widespread installation of the new and improved Westinghouse vault transformers.

Electrical World documented in 1928 the placement of those transformers in the basements of theaters and commercial structures to maximize distribution efficiency. It noted that the United Company had pioneered much of that effort with the comment that "the economic justification of alternating current service has not deterred utilities from seeking reliability equivalent to that of Edison direct current service." It went on to note the pioneer role of United in the substitution of networks in place of single feeder lines to increase reliability.

The simplicity of the United system advanced the design and construction of large buildings which experienced a continual increase in electric power demand. United installed 4,160 volt lines to distribution transformers within the new Columbia Presbyterian Medical Center in 1928. United connected 7,500 volt transmission lines to the distribution transformers in refrigerated storage warehouses on the lower west side. The refrigeration market was one that was sought eagerly by the

VI: ALTERNATING CURRENT INCURSION

United Company as the "leading current" or reactive component produced by the synchronous motors used to drive the refrigeration compressors acted to balance the "lagging current" reactive component consumed by the induction motors in adjacent industrial plants and thus enhanced the overall power factor of the system without additional compensation. Thomas E. Murray was a board member of some of those refrigeration companies.

A new practice was introduced in the construction of the Chrysler Building of 1929. In that structure, the 13,200 volt cables extended directly to transformers in the basement and on upper floors. That created a "vertical" distribution network which was an extension of the horizontal street network perfected with the developments of 1923 and 1926. The use of high voltage reduced the current, and thus the number, of vertical "riser" cables, to the extent that space occupied by cables in previous buildings was available for rental. Adopted for other large buildings, it was introduced to Chicago in the Field building. A similar arrangement was installed in the Federal Archives building in Washington.

Frank W. Smith, Leader of Innovation at United

Frank W. Smith encouraged the development of new concepts at the United Company. Smith began his career in 1880 as a twelve year-old office boy in the United States Illuminating Company, the predecessor of

United. Rising to Secretary by 1905, he was appointed Vice-President in 1912, and added General Manager to that title four years later. Descended from Revolutionary War families and Sir Francis Drake, Smith had an active civic sense and a gift for public relations. Proud of the leadership of his company in the development of alternating current, he encouraged the use of public print to promote the United Company. His effort led to the construction of the new power plants and the transformer stations. He also directed the development of service buildings, offices, and promotional programs.

Matthew Sloan

Another proponent of alternating current was Matthew Sloan. A rising figure in the New York utility industry, Sloan was elected president of New York Edison and also of United in preparation for an eventual merger. Born in Birmingham, Alabama in 1881 and trained as an engineer, Sloan rose through the ranks of electric railways and utility companies in the South. Moving north to New York Edison in 1917, he assisted John W. Lieb and Thomas E. Murray at New York Edison prior to his election as president of Brooklyn Edison. Sloan's understanding of technical principles was rivaled only by that of Lieb and Murray. His business knowledge was on a par with that of Murray and Smith.

Sloan had initiated in 1923 the retirement of direct current distribution in Brooklyn to modernize the

VI: ALTERNATING CURRENT INCURSION

system, reduce expense, and enable all customers to use new products. He had also directed the substitution of three-phase lines in place of the original two-phase system in Brooklyn. His administration had constructed the new Hudson Avenue generating station that set world records. Designed by Murray, it adjoined the pioneer Gold Street plant of 1900 and avoided the acquisition of expensive property for new ducts by an increase in transmission voltage from 13,000 to 27,600 to reduce the number of cables required. As president of New York Edison, Sloan ordered a thorough review of the customer service.

VII: ALTERNATING CURRENT TRIUMPH = DIRECT CURRENT DEMISE

At the time of the First World War, the downtown areas of dense load concentration in most cities were supplied primarily by direct current distribution from substations with battery protection. Change came rapidly in a revolution that swept the United States in the early 1920s. The number of utility companies across the nation that initiated the retirement of direct current distribution was such that it generated a series of extensive articles in *Electrical World* during 1922 and 1923. It was noted that most utility companies were "hard at work" on plans and/or programs to retire direct-current systems.

Louisville, Newark, Omaha, and San Diego had made alternating current the principal system. Los Angeles, San Francisco, and Portland, Oregon had limited the extension of direct-current distribution while

Seattle had retained it only where the 2,300 volt alternating-current transmission system was inadequate to handle downtown loads. Washington D.C. was installing 60 Hz lines in preparation for a future change. Only the Boston Edison, Commonwealth Edison Company of Chicago, Philadelphia Electric, and New York Edison Companies appeared intent on the retention of direct-current distribution in downtown areas. Another article on the advancement of alternating-current distribution acknowledged that the continuity of it was not always the best, but also noted that failures of direct-current substations and cables had been the subject of a recent conference.

Another feature explored the issues of customer satisfaction with company policy based on the experience of the author, F.S. Root, in Fall River, Massachusetts. Beyond the superiority of direct-current motors on elevators and some machinery, it noted that the hum produced by the alternation of the magnetic field in alternating-current motors could be objectionable in some circumstances. Business machines, the cash carrier systems of department stores, and church organ blowers were detailed as examples. It stated that the "change-over" process would have to be undertaken at company expense but warned that advance notice should be withheld.

Although that policy might anger some customers about to purchase new equipment, it was necessary to prevent the unscrupulous from stockpiling old

VII: ALTERNATING CURRENT TRIUMPH = DIRECT CURRENT DEMISE

direct-current equipment. Otherwise, worthless junk might be exchanged for brand new replacements which the customer could then sell at list price. It was suggested for the same reason that all equipment turned in to the company should be destroyed immediately. If otherwise discarded, it had a tendency to reappear in the hands of another customer who would then turn it in again for exchange.

There was also the problem of wiring which sometimes required an upgrade and could become the source of dispute between utility and customer. It was noted that it is natural to remember the best aspects of the past, a tendency which was often manifest in an attitude that the new was inferior to the old. For that and other reasons the author recommended that equipment of only the best quality should be exchanged for obsolete items. The disturbance to business or premises during the changeover process could also become a consideration.

Changeover in the Major Cities

The Commonwealth Edison Company of Chicago initiated in 1923 plans for an eventual retirement of direct current with a program that added 60 Hz distribution to areas served by the old system. The fringe areas of the direct current distribution system were gradually changed to alternating current, effectively shrinking the direct current core of the city. Philadelphia was the first of the four large systems to initiate a total changeover process. The Philadelphia Electric Company announced also in 1923 that the 1.7 square mile downtown direct current district would be converted to alternating-current distribution. That decision was motivated by two primary concerns. First, the construction of substations in a congested city was an expensive proposition. Second, the underground vaults and cable ducts were so crowded that faults on one line often "flashed over" to an adjacent cable. Substitution of high voltage 60 Hz cables would reduce the current to the extent that fewer cables would be required to supply the area.

Customers were offered two options in regard to the replacement of old equipment. The first was a direct exchange of new motors or appliances for old. In the event that a customer preferred to retain the old items, a free replacement would be "loaned" so long as the account was maintained at that address. In practice, the value of the item depreciated over several years and the item effectively became the property of the customer. It was a small price to pay for the retirement of the overloaded and trouble-prone direct-current distribution system.

Direct-current distribution in Brooklyn was limited to the older neighborhoods and commercial areas. There were only eighteen substations, and none had the extensions that were added so frequently in Manhattan. The direct-current system in Brooklyn peaked in 1922 with 125,000 customers and a load of 75,000 kW. Most Brooklyn substations included 60 Hz equipment for new

VII: ALTERNATING CURRENT TRIUMPH = DIRECT CURRENT DEMISE

customers as they were almost invariably supplied with alternating current. The only exception was increased direct-current load in the Wallabout Market area and also from industries around the Navy Yard. That required a new substation, the last extension of the direct- current system in Brooklyn. Brooklyn's alternating-current load surpassed the load on the direct- current system for the first time in 1922. A year later, the company announced that direct-current distribution was to be retired. In order to avoid the expense involved in the replacement of motors, priority was accorded those customers for whom lights were the primary load. Thus residential customers were often the first to be changed. Thus 1923 ended with commitments from three of the largest utility companies (Philadelphia Electric, Commonwealth Edison, and Brooklyn Edison) to retire direct-current distribution.

Five years later, change came to New York Edison. Matthew Sloan released the result of his survey of New York Edison D.C. service in November, 1928. His report stated: "in light of the reliability, economy, and efficiency of the automatic alternating current network perfected by the United Company, there is no justification for continued extension of direct current service." Sloan then issued a series of orders to retire direct current at company expense. That launched one of the most ambitious, extensive, and complex programs ever undertaken by any utility executive. The program, based on his experience in Brooklyn, favored the changeover of those services in which lights were the primary load. However, the cost of replacement motors would be borne if it would prevent the construction of new direct-current substations or extensions to existing facilities.

Sloan's first order, designated T-6 and dated 20 November 1928, restricted the extension of direct-current service. Order T-28, dated 31 December; mandated the changeover of service where lights constituted the primary load. Subsequent orders placed restrictions on the extension of direct-current distribution in those areas where the capacity in place was overloaded; in particular the midtown area supplied by the West and East 39th Street substations. Order T-42, dated 21 February 1929 stated that only alternating current was to be provided to new customers in the area below 59th Street. Two planned direct-current substations and one extension were canceled. New requests for direct current would be provided only where substation capacity was adequate. As lighting loads received priority status, street lights were to be transferred to alternating-current lines immediately. All ducts were to have 60 Hz lines installed as soon as possible.

In 1928, the company reviewed major buildings under construction or renovation and recommended either alternating current, direct current, or in a few cases both, based on the substation capacity available at that location. The list of 1929 was primarily alternating current; the subsequent review of 1930 had no

VII: ALTERNATING CURRENT TRIUMPH = DIRECT CURRENT DEMISE

direct-current recommendation. However, the sheer momentum of established practice continued to increase the direct-current load. That peaked in 1930, despite the transfer of ninety-nine percent of the street lights and some customers to the new lines. The direct-current load in Manhattan had increased 560% between 1910 and 1929, 238% from 1920 to 1929. Sloan estimated that there was about one hundred million dollars worth of investment in direct-current equipment which would be replaced only as it wore out or was superseded by new designs. He stated that it would take forty-five years for the company to retire the old systems completely.

Sloan ordered that the expensive concentric cable (negative line inside a hollow positive lead) used in the direct-current distribution was to be salvaged as the service was withdrawn. Salvaged cable would be stored for use in repairs or in the few additions to the old system. Most of those additions were the substitution of Edison power for private plants. New York Edison continued to replace private plants at the rate of twenty to twenty-five per year.

Most customers appear to have welcomed the changeover. While it may have caused some nuisance, the list of products that could not be operated on direct current had increased beyond small appliances, compact business machines, and clocks. Residential customers could not use a number of new devices. New cooking and heating devices often incorporated thermostats with contacts that could not withstand the arc produced by an interruption of the flow of direct current. Any item powered by a small transformer could not operate on direct current. Those included some medical devices and household items such as electric toys, phonographs with electronic amplification, doorbells, and most radios powered by household power. As an interim measure, some large offices, apartment buildings, and hotels installed separate alternating-current circuits to permit the use of modern devices. The direct-current systems in place were then often retained in order to supply motors. As the receptacles were identical, such an arrangement could create havoc in hotels when visitors plugged an appliance into the wrong line.

New York being what it is, the transition was not without opposition. The Structural Steel Board and other representatives of the construction industry sued; the allegation was that New York Edison had encouraged the use of direct-current motors on construction machinery with the claim that they provided superior performance and control. Others claimed that the use of alternating-current motors was more expensive when compared to direct-current motors in the same application. Heard before the New York State Public Service Commission, the settlement required New York Edison to continue direct current to construction sites so long as the supply was adequate. The Public Service Commission had been established 1 July 1907 to regulate intrastate utilities and transportation companies.

VII: ALTERNATING CURRENT TRIUMPH = DIRECT CURRENT DEMISE

The Passing of Giants

Thomas E. Murray, the man who shaped the electrical system of New York City, died at Wickapogue, his summer home near Southampton, Long Island on 21 July 1929. Overall, he designed nine large power stations in the city. Seven were general service, the other two were constructed by the Brooklyn Rapid Transit Company. He was involved in the design of the direct-current substations and the alternating-current transformer substations. He also designed power plants for industry, government reservations, and cities in upstate New York and in the South. Murray detailed that work in a series of books on the subject of power plants, substations and service buildings.

Murray pioneered the use of automatic boiler stokers and dust catchers for coal storage areas. He developed a water-cooled, steel-walled furnace that was introduced in the Hell Gate station of the United Company that opened in 1921. A new standard in the design of power stations, it saved much of the space occupied by the brick furnace walls of previous plants. In addition to his position as Vice-President and General Manager of New York Edison, Murray held the Vice-Presidencies of the Central Telegraph & Electrical Subway Co., the United Electric Light & Power Co., and the Edison Electric Illuminating Co. of Brooklyn. The latter was reorganized as Brooklyn Edison in 1919.

Beyond the electric utility business, Murray held management positions at major refrigerated warehouse and storage companies. His inventions and promotions fostered the development of dishwashers and household and commercial refrigeration. His promotion of safety features in gas and electric devices improved products that ranged from commercial ovens to household appliances and Christmas tree lights. That concern for safety brought recognition in the form of the 1910 Longstreth Medal of Merit by the Franklin institute in Philadelphia and the 1913 award of the American Museum of Safety. He received an LL.D. from Georgetown University in 1918.

He was granted numerous patents for new electric welding techniques which were applied first to the manufacture of artillery shells during World War I, then to the manufacture of automotive parts, and finally to new and more efficient radiators for large buildings. Overall, he received more than four hundred and sixty patents for his innovations. Declining health forced Murray to resign the chairmanship of New York Edison in November, 1928. He continued work at his consulting firm, Thomas E. Murray Inc. and also at his Metropolitan Engineering Company. He also continued to supervise the Metropolitan Device Co. which manufactured large electrical components such as the reactance coils which he developed to protect alternators from power surges. He continued to direct operations

VII: ALTERNATING CURRENT TRIUMPH = DIRECT CURRENT DEMISE

at the Murray Radiator Co. which manufactured the radiator systems he had patented.

Electrical World eulogized his life: "Rich in the world's goods, rich in progeny and rich also in accomplishment, Mr. Murray rounded out what his intimate friends knew to be a well ordered life. There was no show or braggadocio about him through his allotted three score and ten years. As the industry grew by leaps and bounds, Thomas E. Murray grew with it." The feature went on to note that he had power stations of more than two million kilowatts combined capacity to his credit.

The obituaries that appeared in the local newspapers tended to focus on his inventions but noted that he had effected the creation of the electrical companies and networks which served New York City and part of Westchester County. It was noted that he was held in such high regard that he was placed in virtual charge of all major operating decisions and policies established by the companies. Furthermore, the articles declared that the creativity and genius of Murray brought a uniformity to the systems which facilitated interconnection and eventual consolidation. His efforts in regard to safety in electrical and gas systems and appliances were praised while his charity in church work earned him membership in the Knights of St. Gregory and also in the Knights of Malta. Unlike most tycoons of the day, Murray demonstrated genuine concern for the needs of workers and avoided major labor conflicts.

While the talents and accomplishments of Thomas E. Murray were many, one very impressive characteristic emerges in a study of his work from the vantage point of seventy years. That factor is the equanimity of Murray in his interaction with associates. Murray partnered with John Lieb in the advancement of direct current distribution systems to meet immediate demand. At the same time, he supported strongly the efforts of Frank W. Smith to push the development of alternating current for the future, while his companies developed components such as network protection devices that were vital to alternating-current distribution. Murray's faith in the future dominance of alternating current distribution was exemplified by the initial network transformer protective devices developed and advertized by his company as early as 1913, well before refined network systems had been perfected. Such vision and confidence in the orderly progression of technological development is a rare gift indeed.

John W. Lieb, the man most associated with the development and installation of the gargantuan direct-current distribution system of New York Edison, died in New Rochelle, N.Y. on 1 November 1929. As Vice-President of New York Edison, Lieb had continued to design improvements for that system. He held executive positions in both the American Institute of Electrical Engineers and also in the Association of Edison Illuminating Companies. He remained steadfast in his favor of direct-current distribution in the downtown

VII: ALTERNATING CURRENT TRIUMPH = DIRECT CURRENT DEMISE

areas of large cities. Only at the opening of the East River station in 1926 did he finally acknowledge publicly that the future would likely see standardization of 60 Hz power. Lieb's prized collection of materials relating to the life and works of Leonardo DaVinci was donated to the Stevens Institute of Technology.

VIII: CONSOLIDATION

While the Depression that followed the stock market crash of 1929 reduced revenue, the economic downturn had one positive aspect for electric utilities. The retirement of direct-current systems advanced as the load declined. The process was further expedited by a merger of the New York Edison technical department with that of United in 1932. That made possible the direct transfer to New York Edison of more than thirty years of United Company experience with alternating-current systems in urban areas of dense load concentration. (Both companies had moved their headquarters into the same office building in 1914.) The two companies merged to form the New York Edison Co. Inc. in 1935. A year later, that company and a small Bronx affiliate were merged with Consolidated Gas to create the Consolidated Edison Company Inc. Brooklyn Edison and the NY & Queens Co. were unified over the next decade to complete the system envisioned by Murray more than four decades earlier. The Yonkers and the

Westchester affiliated companies were merged in the early 1950s while the independent Staten Island Edison Company merged in 1950.

Customer Service Changeover

New York Edison had continued to experience a substantial decline in direct-current load after the worst part of the Depression ended. New buildings were completed and others renovated while some customers were changed in accordance with the priorities established by Sloan. In 1928, the last full year included in the Sloan report, the total distribution capacity in all of Manhattan was 524,000 kW. That was comprised of eighty-four percent New York Edison Co. direct-current distribution (441,000 kW.) and only sixteen percent United Co. alternating-current distribution (83,000 kW.) Five years later, the direct-current load had declined to the extent that batteries were removed from four substations, and one was closed at night.

By 1937, more than 19,000 New York Edison customers had been changed and several substations in business and commercial districts were closed at night. One of those was the West 22nd Street substation in the heavily loaded midtown area where Edison had built his first Vertical stations in 1888. Manhattan distribution capacity of the combined companies in 1937 totaled 654,000 kW. which was comprised of forty-eight percent direct current (317,000 kW.) and fifty-two percent

VIII: CONSOLIDATION

alternating current (337,000 kW.) In summary, direct current load had been reduced twenty-eight percent (124,000 kW.) but alternating current load increased 306% (254,000 kW.) Overall, new alternating current load totaled 130,000 kW, or 156% of the 1928 United system capacity.

As the economy continued to improve, new construction and the modernization of older structures accelerated the change. In some areas, the demolition of entire blocks of old buildings produced a steady decline in the direct-current load. Modernization of factories, bakeries, commercial kitchens, and department stores reduced the load on substations throughout Manhattan. One of the largest was the kitchen and bakery of the Horn & Hardart company, the owner of the famous Automat restaurants. It occupied an entire block on the west side and processed freight car loads of food daily. The modernization and new equipment was such that the load on adjacent substations was cut by nearly forty percent. By 1943, the capacity of direct current substations throughout Manhattan was reduced sixteen percent, as the load had declined thirty percent. The number of rotary converters had been reduced fifteen percent and all of the batteries were removed.

The decline in direct-current load reduced substantially the load on the 25 Hz generation and transmission system that supplied the substations. The changeover required adequate 60 Hz capacity to meet the demand of those customers changed to alternating current. That demand was supplied initially by the United Company from capacity at its Hell Gate station in the Bronx. However, both Sherman Creek and Hell Gate also supplied a substantial amount of 60 Hz power to the substations of the new Independent Subway System that opened in stages between 1932 and 1940.

As in the case of previous railways, a power station had been planned for the new subway. A site was selected north of the Waterside complex but construction bids exceeded estimates by 400%. At that point, it was decided to use power supplied by the utility companies, a trend which had developed throughout the railway industry. The result was a twenty-year contract to supply power to the substations of the new subway line. Given the advanced state of 60 Hz systems, it was decided that the substations would operate on power of that frequency supplied from stations of the United and the Brooklyn Edison Companies. New York Edison was listed as the utility; a similar contract had been issued in 1914 when the United Company installed a limited 25 Hz system in the Sherman Creek station to supply the New York, New Haven & Hartford Railroad. Regardless of the paperwork, the load of the subway accounted for a large proportion of the available 60 Hz reserve. By 1937, the reduction of the Edison direct current substation load was such that the 25 Hz generation and transmission capacity exceeded demand by a significant margin.

As a result, Waterside Station Number Two was converted to supply the 60 Hz load of the area between

VIII: CONSOLIDATION

14th and 59th Streets. Yet another massive renovation program was undertaken to modernize the station. The ninety-two boilers installed during the previous modernization were replaced with eight high capacity units. Topping cycle turbines were added to use the low pressure steam that was exhausted from the large turbines. Electrical modification involved alternators, transformers, controls and protective equipment. By the end of 1938, Waterside Number Two supplied much of the alternating current consumed by midtown customers. That released a substantial amount of the 60 Hz capacity of United to serve new alternating-current customers in other areas, especially those in new residential and commercial structures in lower Manhattan and on both sides of Central Park.

The 25 Hz load of the direct current substations and the railways was supplied by the capacity in place and that of the semi-retired Brooklyn Edison Gold Street station. Reopened after having been placed on reserve during the worst part of the Depression, it was operated to meet peak winter demand as needed. (The 1897 vintage Bay Ridge station had been idled due to obsolescence.) In 1938, the Brooklyn Gold Street station was linked to the Hudson Avenue station by steam lines to permit a rapid start without the delay required to warm cold boilers. At that time, Consolidated Edison supplied the 25 Hz load of the substations of the Long Island Rail Road with the acquisition of a railroad power plant in Long Island City, Queens.

As in the case of the other rail lines, the railroad had determined that it would be more economical to purchase power than to continue to operate the station as a base load plant. The railroad had purchased 25 Hz power from Brooklyn Edison for some twenty years whenever the peak load surpassed the capacity of the Long Island City plant. By 1938, the railroad imposed the bulk of the 25 Hz. load on the Brooklyn Edison system for the changeover of direct current customers to alternating current that had been initiated in 1923 progressed rapidly. By 1938 there were only three substations needed to supply four hundred and fifty direct-current customers with a maximum 7,000 kW load. Overall the number of Brooklyn Edison direct-current customers had been reduced by 99.996%, the substations by 83% and the peak load by 99.994%. Those remaining Brooklyn customers appear to have been primarily industrial firms along the waterfront and commercial buildings in downtown Brooklyn.

Change in Midtown

The West 26th Street substation continued to supply the heavy direct-current load of motors and lights in the loft buildings of the garment district and fur trades. That increased as the economy improved in the late 1930s. In 1940, the substation still supplied a total of one hundred sixty-nine feeder cables, more than any other. The redevelopment anticipated after the 1938

VIII: CONSOLIDATION

closure and subsequent demolition of the Sixth Avenue elevated railway did not occur, and the loft buildings remained. Nor did the December 1940 opening of the new Sixth Avenue subway produce significant redevelopment.

The opposite was the case several blocks to the west. Modernization of warehouses near the Hudson River and the construction of new apartment buildings on the lower west side reduced the demand on the West 27th Street substation to the extent that only one rotary converter was in operation. As the site was selected by the city for a park and housing project, it was determined that other substations had sufficient capacity to assume that load. Most of the load was transferred to the West 36th Street substation. The balance was assumed by the West 22nd Street substation and a more distant facility on Horatio Street. On 11 May 1941, the last feeder was transferred and the substation closed, the first direct current substation to be retired.

The load on the West 39th Street substation was reduced by the modernization of large department stores such as Lord & Taylor. Still, the demand imposed by the women's fashion and millinery trade, textile businesses, large stores, and the Times Square area continued to make it the most heavily loaded substation on the Edison system. It retained the record for substation capacity and also for hours of heavy load. The latter was imposed by the theaters, restaurants, and older hotels in Times Square.

Old #9, Star at the World's Fair

Advertisements and related programs declined with the Depression, but increased as the economy recovered. The most significant company promotion of the 1930s was an exhibit at the New York World's Fair of 1939-40. Detailed miniatures of Waterside and other plants were part of an extensive Consolidated Edison display. The star of the show was the old generator #9 from Edison's original Pearl Street station. The only unit to survive the disastrous fire of 1890, it had been retired with the station. After years in storage, it had been displayed at the 1904 St. Louis World's Fair. It was then stored at the Shadyside coal storage bunkers on the west bank of the Hudson River. Located opposite West 96th Street the bunkers maintained coal reserves for the generating stations of both United and N.Y. Edison.

Overhauled for a 1932 exhibit that marked the fiftieth anniversary of the Pearl Street station, the old unit was stored again until the Fair. It was then restarted to supply some of the power for a television special in which it was the star. That event on 11 September 1939, was broadcast from 8 P.M. to 9 P.M. by the experimental NBC television station W2XBS which had initiated commercial television broadcasting at the opening of the Fair. Ultimately # 9 joined components of the old Duane Street Vertical station at Henry Ford's Greenfield Village exhibit near Detroit MI.

VIII: CONSOLIDATION

Social Intervention

The 1930s marked a turning point in the electric utility industry in much of the nation. Popular opposition to large businesses became widespread during the Depression. Critics noted that the thirty year period from 1902 to 1932 had been marked by two trends. The first was an increase in the number of electric utility companies from 3,620 to more than 6,000. The second was a decline to 3,400 as a result of mergers. As a result the fifteen largest firms controlled thirty-five percent of the generating capacity. That was to be expected, for it was a period when urban expansion was most rapid.

Serious study would have revealed the simple fact that merger and rationalization was the only means by which large scale electrification could be capitalized. Electrification made possible many improvements in the life of the average citizen, improvements which had made the United States the envy of the world. Nevertheless, the charges of improprieties and abuses were rampant. It was alleged that New York Governor Franklin Roosevelt forced out Public Service Commissioner Prendergast as being too "utility friendly."

Nationally, the controversy produced the Public Utility Holding Companies Act of 1935, and also created the Federal Power Commission. The federal programs of the "New Deal" of President Franklin Roosevelt included rural electrification, hydroelectric dams, and reclamation projects. Developed areas in which large scale electrification was in place did not experience radical change. For New York City and Consolidated Edison, that period was marked by continuance of the projects initiated prior to the Depression.

World War II

The advent of World War II produced a substantial increase in the load on the Consolidated Edison system. Industrial load increased on both the alternating and the direct-current distribution systems. Domestic load increased around the clock as war-related industries went to three shift schedules. The transportation load increased as rationing of gasoline and rubber put more passengers on railroads and transit lines.

Some direct-current substations in the industrial areas of lower Manhattan had supplied only a minimal load after factory production declined during the Depression. Defense mobilization produced a flurry of new business and those substations were suddenly called upon to supply full power around the clock. However, the overall 25 Hz load continued to decline through 1944 as direct-current customers were changed to alternating current by renovation and construction projects that had been initiated prior to the war.

IX: POSTWAR PROSPERITY

Once hostilities had ended, the focus of utility development shifted to the revival of expansion plans that had been deferred by the Depression and the war. Two additional stations were designed, and new transmission and distribution concepts were developed to supply power to new and renovated buildings in Manhattan. The load on the direct-current substations declined to the extent that the company was able to reduce the 25 Hz generation and transmission capacity once again. One of the former New York Central Railroad power plants (Port Morris in the Bronx) and that of the Long Island Rail Road were retired. The only factor that offset the rapid decline in direct-current load was the continued retirement of private plants in favor of utility power. Several large department stores and commercial buildings preferred to retain direct-current systems to permit a gradual renovation of their premises. Very few commercial customers intended to retain direct-current systems more than a few years. Therefore the overall number of new direct-current customers was minimal.

Passing of the Last Pioneer

Frank W. Smith was the only pioneer to see the new postwar era of urban electric service. The leading business figure in the program that established 60 Hz as standard in the dense urban conditions of Manhattan, Smith died on 22 July 1946 at the age of seventy-nine in his home at 277 Park Avenue. He had remained active in the business long after the final triumph by United in 1926. Elected vice-president of New York Edison in 1931 in preparation for the merger with United, Smith was elected president of both companies in 1932. Three years later he was elected president of Consolidated Gas and the New York & Queens Electric Light & Power Company. Upon the merger that created the Consolidated Edison Company, Smith was elected president. He was also elected president of Brooklyn Edison which remained independent at that time.

Smith retired in 1937 but remained active as a director until 1945. He held major positions in most of the electric companies in the city at various times. He also held a number of honorary offices and served on various industry associations and managed employee benevolent funds. As United pursued the technical goals of George Westinghouse, so Smith promoted at United the development of employee protection plans similar

IX: POSTWAR PROSPERITY

to those initiated at the Westinghouse Company. Those plans included pension, disability and survivor insurance. A trustee of hotel and insurance companies as well, he maintained an interest in the city and civic life. An enthusiast of the sport of "automobiling," Smith was the coauthor of the *Electric Vehicle Handbook* of 1913.

Although the research and inventions of Lamme, Murray, Stanley, Shallenberger, Steinmetz, Tesla, Thomson, Westinghouse, and others made modern alternating-current systems possible, the contributions of men like Smith are often overlooked. His direction and dedication placed the United Company at the very forefront of the utilities that pursued the development of alternating-current systems. The success of that effort demonstrated to the industry that alternating current could not only accommodate any load but also that it had come of age for any locale.

His obituary in the New York Times noted that he had "brought about" the new generating station (Sherman Creek) and transformer stations constructed by United prior to World War I. Since the company was controlled by Consolidated Gas at that time, it may be assumed that it was his efforts in management and finance that were referenced. Not as famous as either Murray or Lieb, Smith was survived by a wife and daughter. His quiet sense of humor was detailed in an impromptu harmonica concert and contest in Bryant Park. He and utility engineer Charles Powell joined former Governor Alfred E. Smith, much to the delight of observers who knew only his conservative approach to business.

New Development Brings New Loads

With the Depression and World War II over, the United States enjoyed one of the longest periods of sustained economic growth in the history of the world. Manhattan was filled with new commercial and residential properties while the outer regions of Brooklyn, Queens and the Bronx developed as suburban residential areas. Existing structures were modernized to accommodate the flood of new appliances and machines that were entering the market. Furthermore, the very appearance of urban space was changing. Consolidated Edison greeted the end of the war with a special light bulb promotion and the slogan "Lights On" to signal the end of wartime restrictions. That was just the beginning.

Bright illumination of bright colors and chrome was the defining motif. Fluorescent lamps were the new standard for stores and offices. The use of incandescent lamps to highlight areas was replaced by fluorescent lamps for general interior illumination, while gas discharge "neon" signs were the new standard for advertising. Small incandescent lamps and/or spot lights marked a proprietor as being out of step with the new postwar world. All that brightness imposed an increased electrical demand.

IX: POSTWAR PROSPERITY

Household appliances had been limited to a maximum of 660 watts in the 1920s, a restriction determined by the lamp sockets which had been used in some instances to supply power. Improved standards removed those limits. By the 1950s, cooking and heating appliances often exceeded 1,000 watts; some were rated as high as 1,600 watts. Television receivers were a new residential load. The sets of the period consumed a substantial amount of power, often 400 watts or more.

The actual and relative cost of electric power continued to decline as the increased sales combined with technical advance to produce greater economies of scale. In the 1950s, the average cost of residential electric service nationally was little more than three cents per kilowatt hour. Expansion of the economy released consumer demand for new products that had been constrained first by the Depression and then by the War. Electric industry trade organizations promoted increased power consumption with the slogan "Live Better Electrically."

To so live required expansion of electric capacity throughout the nation. While the postwar boom continued, an underlying fear of resumed global conflict persisted. As the Cold War deepened, the executives of electrical manufacturing companies such as General Electric assured the public that their firms could supply adequate generating capacity in the event of renewed hostilities. Consolidated Edison advertisements promoted new products and detailed the construction of facilities. Public tours of the Waterside station which had been suspended during World War II were resumed.

Expansion

Consolidated Edison initiated a massive expansion program that dwarfed the efforts of the individual companies of previous decades. Sixty Hz generating stations were modernized to increase capacity. Additional capacity was developed through the introduction of sealed alternators that were pressure-cooled by hydrogen gas. When large 25 Hz alternators were rebuilt to supply 60 Hz power, the substitution of the pressure cooling system increased the rating twenty percent from 160,000 kW. to 200,000 kW.

New stations in Queens and Staten Island contained alternators that were rated at 300,000 to 400,000 kW, as much power as the entire Waterside complex produced in the 1920s. The program marked the return of generating capacity to Queens. Staten Island was a region which was developed so sparsely that the Livingston station of 1896 remained the sole source of power through World War II. It was only retired in 1961 after new generation and transmission capacity was completed.

The Waterside complex continued to keep pace with new developments. At that time, it supplied the most dense load of the city, the territory that included the Empire State Building and the East 42nd Street area

IX: POSTWAR PROSPERITY

which had experienced the most rapid redevelopment. The combined capacity of the Waterside stations reached 658,000 kW in 1949. At that time, the plants contained fourteen large turbine alternators; five supplied 25 Hz power, the balance were 60 Hz units.

Improvement of the transmission system was required to control and regulate the power produced by that massive increase in generating station capacity. The load in midtown Manhattan increased to the extent that 13,000 volt transmission cables crowded underground ducts. Distribution system efficiency and capacity were improved substantially by the introduction of high voltage lines that increased capacity by as much as 1000%. The number and bulk of cables was reduced accordingly. Still, additional duct space was required and an extensive expansion program in the 1950s produced a new slogan at street construction sites: "Dig We Must For A Growing New York."

That program often encountered obstacles as the underground space in midtown Manhattan is as crowded as the space above, and any available space was used. In one instance, an ideal solution was found in the conduits which had been constructed for nineteenth century cable cars. Power rails had been installed subsequently but removed upon the termination of streetcar operation. Because the space was intact, the installation of electric cables was accomplished with little disruption to street traffic, much to the delight of police, traffic officials, motorists, and pedestrians.

Though not part of the Edison system, the lines within large buildings evolved to meet new demand more efficiently. Building networks energized at 265/460 volts and later 277/480 volts supplanted the standard 120/208 volt lines that were introduced with the three-phase networks of United in the 1920s. One of the first of those new network systems was installed in the printing plant of the *Daily News* in the early 1950s.

New Air Conditioning Tradition

Widespread adoption of air conditioning, both central and unitized, produced a drastic increase in the electric load on the system. Since the earliest days of the business, the peak demand experienced by urban electric utilities had always occurred in late December and early January. The apex of that peak was reached around 5pm when the short days produced a heavy demand for illumination at a time when a large number of elevators and industrial motors were still in operation. With the widespread installation of air conditioning in the late 1950s, a new summer peak developed on the afternoons of hot summer days. Tradition was reversed on 10 June 1959 when the peak load on the Consolidated Edison system surpassed that of the December, 1958 - January, 1959 winter peak. Thereafter, the peak demand has always occurred on a hot afternoon in the summer.

IX: POSTWAR PROSPERITY

Milestone Year: 1962

The eightieth anniversary of Edison's first station on Pearl Street was marked by major developments. The office of the system operator was replaced to accommodate the complexities of the consolidated companies. A scheme to end the direct current substations proved successful. Additional capacity was provided by the first nuclear power station in the area.

The retirement of the direct current substations was initiated after a 1961 survey of the direct current customers determined that the efficiency of the system could be improved substantially. Twenty three substations remained in operation as the retirement rate averaged only one a year. Many of those supplied a minimal load but could not be retired because of the short range of 120-volt direct-current distribution. New developments in solid-state electronic technology promised a solution in the form of a "rectifier" that required no supervision and could operate on the standard 60 Hz alternating-current system. It was compact and could be located near the load in order to eliminate long cable runs.

The rectifier concept had been explored by New York Edison in the 1920s with an experimental installation of mercury arc rectifiers at the Gold Street substation in lower Manhattan. Not to be confused with the Gold Street power station of Brooklyn, the facility in Manhattan started life as an arc light station of the

Excelsior Company. Though converted to a substation in stages early in the 20th century, it retained stand-by steam boilers and direct-current generators until 1922. After conversion to a substation, an experimental "Dynamotor" converter was installed. Intended to eliminate the need for substation transformers, the dyna-motor was essentially a rotary converter that was powered by three-phase 25 Hz 6,600 volt alternating current directly and produced 120/240 direct current. Insulation was a problem, and short circuits led to a retirement of the scheme and a focus on electronic rectification.

The rectifiers of 1925 consisted of a set of water-cooled steel tanks in which an arc sustained between an electrode and a pool of mercury conducted current in one direction only. Control from a central location was simple because the units required little attention. Efficient in any size, rectifiers could be located close to the load to reduce substantially the voltage drop in the direct-current distribution system. An article in *Electrical World* depicted a rectifier supplied by the Brown-Boveri Company of Switzerland. There is little record of the test, and rotary converters remained the standard for Edison substations.

No specific reason for the retention of rotary converters has so far come to light, however there were several points in favor of the continuance of standard practice. The rotary converters of that time were rated several times the maximum output of the rectifiers then available. Little was known about the long-term operating characteristics and cost of rectifiers. Finally, the capacity of rectifiers of that time was such that a change would have required almost the same substation space. Rectifiers were installed by industrial plants and railways, the aforementioned Independent Subway System used numerous rectifiers in small substations along the lines. Finally the 1928 declaration of Sloan to retire direct-current distribution effectively obviated further consideration of rectifiers.

The development of solid-state electronics after World War II produced silicon rectifiers that were small, compact, and required minimal cooling. Proven in operation on board locomotives and commuter rail cars, the silicon rectifier had reached a state in which it could be placed in underground vaults. Brooklyn was selected for the first installation because only one substation remained and it supplied a minimal number of direct-current customers with less than 1,000 kW of power. Seven small sealed rectifier units rated at 250 kW were placed in underground vaults (apparently located both under streets and also in building vault space) to supply the limited number of direct current customers in downtown Brooklyn. About the size of medium-sized distribution transformers of the ac network, the units measured about eight feet in length, three feet in width and less than five feet in height. The installation proved successful and the last direct-current substation in Brooklyn was retired in early 1963 and with it the 25 Hz Gold Street power station (Murray's first Brooklyn plant) which was demolished in early 1964. Similar installations were planned for Manhattan.

Substantial savings could be realized by the elimination of substation crews, as the rectifiers required only periodic inspection. Plans were initiated to transfer the direct-current load that remained in Manhattan to small efficient rectifiers and close large underutilized substations. In time, the entire 25 Hz system that supplied the direct-current substations could be retired. As the program advanced, the entire generation and transmission system could be revised for greater efficiency as Consolidated Edison acquired three large power stations in 1959. Those plants (59th and 74th Street in Manhattan, Kent Avenue in Brooklyn) supplied 25 Hz power to the rapid transit lines constructed for operation by private companies prior to the Independent System of the 1930s. Constructed between 1901 and 1906 but modernized in later years, the Manhattan plants could supply 60 Hz power for general service. As of 1961, the pooled 25 Hz system capacity was sufficient to permit retirement of generation at the former New York Central Glenwood generating station in Yonkers.

The Nuclear Age

Nuclear generation of electric power had been a topic of serious discussion for many years when the program directed by Admiral Hyman Rickover produced the world's first nuclear-powered submarine, the *U.S.S.*

IX: POSTWAR PROSPERITY

Nautilus. At that time, most utility executives and engineers were focusing on the increased demand for electric power after World War II. Furthermore, many utility leaders were frustrated by the security policies of the Atomic Energy Commission which restricted access to nuclear-related data. One of the most influential executives was Thomas E. Murray Jr., the son of the primary figure in the development of large scale electric utility operations in New York City. A graduate of the Sheffield School of Engineering at Yale University, Murray Jr. had held executive positions at New York Edison and also at the Murray companies.

Named to the boards of local colleges and philanthropic agencies, he was appointed Federal Receiver of the bankrupt Interborough Rapid Transit Company. During a period of eight years he improved subway service and resolved financial and labor problems prior to a takeover of the IRT by the city in 1940. Murray Jr.'s interest in the potential of nuclear power had been piqued by a conversation with a physicist in regard to the future of the subway. As the first engineer appointed to the U.S. Atomic Energy Commission established by President Harry S. Truman, Murray Jr. had extensive knowledge of atomic reactor development and sought to balance the destructive and constructive aspects of nuclear physics. Murray Jr. developed extensive opinion on weaponry and related moral issues as he surveyed research laboratories and observed bomb tests on land, at sea, and in the air.

Murray Jr. was one of the primary figures in the expansion of research through the creation of the Lawrence-Livermore laboratory in California. He explained his opinions extensively and in detail in a 1960 book, *Nuclear Policy for War and Peace*. His discontent with secrecy that he deemed excessive had been echoed by other influential figures, and their combined efforts encouraged a review of that policy in regard to commercial development of nuclear systems. Parallel to the nuclear program of the U.S. Navy, an "Atoms for Peace" program was initiated at the direction of President Dwight Eisenhower. Once that program was in place, Murray Jr. used his influence to have Admiral Rickover direct the development of a civilian nuclear reactor for the generation of electric power. Murray Jr. warned that excessive idealism with regard to nuclear power clouded the issues. He deplored as misleading and harmful to serious discussion the use of catch phrases such as "the atom will make power too cheap to meter." He noted that there was much to be resolved and that substantial investment was required.

The very nature of the effort required a concerted effort by government, electrical manufacturers, and utility companies. Rickover, working with Westinghouse and other industry leaders, adapted a reactor designed for aircraft carrier operation into one which could be used at an experimental power station at Shippingport, Pennsylvania. Intended to supply steam for propulsion turbines on board a ship, it was altered to power turbine

IX: POSTWAR PROSPERITY

alternators. The station produced 60,000 kW, a small amount in comparison to the alternators of Consolidated Edison stations. Nevertheless, it supplied power to the Pittsburgh area through the lines of the Duquesne Power and Light Company. As a demonstration of the potential for the practical application of atomic physics, the station encouraged subsequent projects.

There were many obstacles to the translation of a military program to the civilian power industry. Later accounts indicate that, like the purchase of Eickemeyer by General Electric to obtain the services of Steinmetz, the most significant transfer was people rather than hardware. Many of the key figures in the development of nuclear power stations came from the Navy program directed by Rickover. Executives of the major electric utilities initiated programs to develop nuclear power stations. Consolidated Edison, Commonwealth Edison of Chicago and the Philadelphia Electric Company were among the first utilities to commence construction. The New York utility owned land on the Hudson River near the town of Buchanan. Named Indian Point, it was selected for the site of the first Consolidated Edison nuclear generating station.

After extensive design and licensing review, the plant was approved and construction was initiated. On 2 August 1962, the reactor "went critical" as the first nuclear reaction was sustained. On 16 September, nuclear-generated electric power was sent to New York City, eighty years after Thomas Edison and John Lieb initiated operations at Pearl Street. Consolidated Edison led an exclusive club of nuclear utilities. Nuclear generation held a particular appeal for the company. Complaints of coal smoke and dust predated electric stations, and those complaints had gained volume with the construction of large stations such as Waterside.

Nuclear generation promised to eliminate even the light smoke of oil-fired plants. The company planned additional units, but encountered public opposition to a proposal to build a unit in Queens. That situation was exacerbated by a statement from one member of the Atomic Energy Commission to the effect that he would not live in Queens if the plant were constructed. Additional units were planned for Indian Point and other locations. The second unit at Indian Point went into operation in 1974. One of the most successful anywhere, it operated at high efficiency and the cost of construction was recovered in three years.

X: RECENT CHANGE AND DEVELOPMENTS

The opening of the Indian Point 2 nuclear station was paralleled by the addition of conventional generating capacity upstate that was shared with other utilities. That permitted the retirement of both the Sherman Creek and the Hell Gate stations in the early 1970s, as well as major reductions in the operation of Waterside and other plants. It also allowed rebuilding of equipment in Queens.

Modernization of the system in midtown included new equipment and lines and the continued retirement of direct current substations. On 24 February 1964, vault rectifiers began assuming the remaining load on Hudson Street on the west side of Manhattan. The program expanded throughout the city over the next decade, and by 1977 only two substations were in full-time operation. As might be expected, they were on the site of the original Edison twins on West 26th Street and West 39th Street.

The 26th Street substation supported 3,950 kW of load which was transferred gradually to rectifiers. With appropriate ceremony, the last load was taken off the 26th Street substation on 3 October 1977. The substation was probably chosen for the ceremony because it was the only remaining structure that dated to the Edison days. The actual last substation to close was the West 39th Street substation, the original Edison structure having been replaced by a typical Murray-era building in 1915. A total 4,500 kW load was transferred to street vault rectifiers by the end of 1977 to end the era of mechanical generation/conversion of direct current. It was proposed as a museum which would have been a great educational tool for both historical and technical subjects. The plans for the museum fell through, allegedly because of political disagreements regarding taxes.

As was the case with all the other substations, it was sold after operations ceased. It was demolished, as many of the others had been. Others, including those on West 48th and West 47th Streets, West 36th and West 35th Streets, East 26th Street, West 22nd Street, West 16th and 17th Streets, and East 12th Street survive, having been redeveloped as commercial space, residences, offices and a health club. At the end of 1977 there were still some 12,000 direct-current customers in Manhattan

Despite the statement of Matthew Sloan in 1928, direct-current distribution lasted much longer than forty-five years. The remaining direct-current customers

X: RECENT CHANGE AND DEVELOPMENTS

hung on— often elevators were the only direct-current load in otherwise modernized structures. Even as the original United alternating-current distribution networks were divided, modified, and modernized to meet the ever-growing demand for power, the remnants of the old direct-current distribution system remained beneath the streets of Manhattan with the street vault rectifiers located to maximize efficiency. The system was costly but the 1929 settlement with the Structural Steel Board was paralleled by an understanding reached with the Public Service Commission—direct current would remain available so long as the customer account remained active.

Finally in the late 1990s an understanding was reached that direct current was obsolete and the utility could charge the true cost of the maintenance of an obsolete system. About 5,000 customers remained on the system in 2000, but soon after that the utility began working with them to arrange either a change to alternating current equipment or the installation of rectifiers on the premises of those customers who wished to continue to operate direct-current equipment. The end came at 2:42 pm on the afternoon of 14 November 2007 at a street vault on the south side of East 40th Street in front of #10 East 40th Street. The severing of the last cable marked the end of 125 years, two months, nine days, 23 hours and 42 minutes of direct-current power sales since Thomas Edison closed the switches at Pearl Street. It likewise marked the final triumph of alternating-current distribution as promulgated by Sprague, Tesla, Thomson, and Westinghouse twelve decades previous.

Little notice was taken of the event because much change had overtaken the business since the great expansion of the 1960s. Despite the substitution of oil for coal to reduce dirt and pollution; then the subsequent adoption of low-sulfur oil, and finally of natural gas, objections to power generation facilities continued. Those objections, combined with the heavy taxation imposed on plants in the city, discouraged additional construction. Plants outside the city were not a simple alternative because transmission capacity was limited, as was the land for additional lines.

In the short term, large numbers of small gas turbine generators were installed, although the scheme was originally developed in the 1960s to provide voltage support during instances of system disturbances, rather than for peak load operation. Though not foreseen as a long-term solution, reliance on such sources of power during summer peak loads became more common as the power business was altered by the concept of deregulation; a process in which utilities were required to sell their generation to outside operators and bid for power. Based on business theory developed from experience with the transportation and communications industries, the final outcome is still uncertain as electric power is instantaneous and the engineering considerations are different.

X: RECENT CHANGE AND DEVELOPMENTS

Power plants themselves changed as large gas turbine alternators took the place of steam turbines. Small gas turbines and diesel generator sets have become a common means to assure that a station has "Blackstart" capability, that is the provision of power to restart a facility that has become an "island" (isolated) after a total shutdown of the system. The nuclear option faded in the northeast despite the successful operation of Indian Point #2 and the additional #3 which was sold to the state power authority before operation commenced. In both instances, the plants paid for themselves in short order and proved to be a reliable source of base load power to the region. The Waterside stations have been demolished and the other plants of the early 20th century have been demolished or are held in a stand-by status.

Furthermore, the century-old annual increase in load faded as equipment manufacturers began a drive for efficiency. Initiated by the double-digit increase in power and fuel costs after the Oil Embargos of 1973 and 1979, the change resulted in dramatically reduced power consumption in commercial, residential, and industrial equipment. As savings were realized, the search for ever greater efficiency accelerated. That alone acted to reduce the need for expanded facilities in the 1980s and early 1990s.

Power transmission over long distances has continued to evolve. The standard approach for most of the 20th century required the planning and construction of new lines based on the estimates of future load. Over time, the land available for such lines was reduced by suburban development. At the same time, new technologies emerged that were made possible by solid state power switching systems. Once again, there was a direct-current option and an alternating-current option.

As noted in Chapter Two, the concept of high-voltage direct-current transmission had been explored in the 1880s. Marcel Duprez transmitted direct current at 2,000 volts in France in 1882. Rene Thury developed a high-voltage direct-current transmission system which was installed in a number of European nations in the 1890s and early 1900s. The Thury system was the original choice of the Cataract Commission for the power station at Niagara Falls until that opinion was reversed by the Westinghouse exhibit at the 1893 Columbian Exposition in Chicago. In time, the Thury installations in Europe were replaced by alternating-current systems based on transformers. Nonetheless, direct-current transmission at high voltage had proponents because there is no loss from those factors which afflict alternating current. Various studies were conducted in England, Germany, and the United States prior to World War II, most of them based on electronic switching with mercury tubes rather than mechanical devices as used by Duprez and Thury.

In the late 1930s, Nikola Tesla, the man whose name is equated with alternating current, told his biographer, John J. O'Neill, that the most practical means by which all the systems in the mainland United States could be

X: RECENT CHANGE AND DEVELOPMENTS

interconnected would use high-voltage direct-current transmission. Tesla noted that substantial progress had been made overseas. Progress was delayed by the war but in the 1950s, high-voltage direct-current transmission was installed in northern Europe and later in areas of long underwater cable runs where it offered significant advantages.

The close proximity of the circuits in an underwater cable increases an electrostatic (capacitance) component of the conductors which imposes a "charging current" which limits transmission capacity. The same is true to a lesser extent in very long transmission lines. High-voltage direct-current transmission obviates those issues but requires expensive and complex conversion stations at each end of the line. Such a line cannot be easily "tapped" along the route to supply intermediate needs. Nonetheless, it has found substantial application in both long-distance and underwater transmission and also in linking systems of different frequencies or those that are not synchronized. Solid-state electronic switching systems that replaced the old mercury arc tubes were introduced in the 1970s and continue to evolve.

The new alternating-current systems take a different approach. As mentioned in chapter three, alternating-current systems are subject to limitations as well, but high-power electronic switching systems can address some of those limitations. Termed FACTS or Flexible Alternating Current Transmission Systems, the new techniques modify power flow as needed to maximize the transmission of useful power as the voltage and current levels in a transmission line change. At the same time they can be operated strategically to control anomalies that occur in complex transmission lines. Such anomalies began to appear in the 1960s when the simple "straight line" transmission of earlier years was interconnected with new circuits. The result was some unanticipated effects such as "loop" flow through adjacent lines or, on at least one occasion, a divided flow through two parallel lines to a distant interconnection.

While there is much discussion and exploration of various "renewable" schemes as well as that of co-generation (in which the "waste" heat is used for other purposes), it remains to be seen whether such schemes will prove permanent or just fads of an era. Those installations are often funded by grants or tax incentives and have been installed by a number of large customers. It is questionable whether those customers will continue to operate those facilities when the full cost of operation and overhaul must be borne by the user as the plants age. While it is much too early to speculate, the lessons of history are obvious. Manhattan once hosted nearly eight hundred private or isolated stations. Then, when the cost of repair or replacement exceeded that of utility power, private "isolated" plants were retired. In a city where heat, hot water, and even air conditioning are in many cases driven by purchased steam rather than individual systems, it would seem that the logical conclusion is foregone.

X: RECENT CHANGE AND DEVELOPMENTS

Regardless of power source, the midtown Manhattan area remains the most dense concentration of electrical load anywhere. Present construction and redevelopment plans appear to assure that it will continue to hold that record. So long as there is a midtown Manhattan like that of the past century and a quarter, the provision of adequate electric power to the area will remain the engineering, operational, and financial management challenge that it has presented since the pioneers sought that market. As has been the case for so long, the systems and technology will have to change and advance to meet that challenge and assure that the provision of reliable electric power will continue.

BIBLIOGRAPHY

<u>BOOKS</u>:

American Committee on Inductive Coordination. Bibliography on Inductive Coordination, (Complete to January 1, 1925) New York, 1925.

Arent, Leonora. Electric Franchises in New York. New York, 1919.

Bailey, Vernon Howe. Electrical New York as seen by Vernon Howe Bailey. New York N.Y.: New York Edison Company, 1916

Beckhard, Arthur J. Electrical Genius Nikola Tesla. New York: Messner, 1959

Cheney, Margaret. Tesla, Man Out of Time. Englewood Cliffs, N.J.: Prentice-Hall, 1981

City Bank Farmers Trust Co. Trust Agreement. The Farmers Loan and Trust Company with Edison General Electric Company and Edward H. Johnson and Frank J. Sprague. December 26, 1889. New York, 1889

Consolidated Edison Co. of N.Y. Inc. An Introduction to Con Edison. New York, 1949

_____. Economic Research Dept. The Trend of Population Growth in Large Cities and their Metropolitan Districts. New York, 1947

_____. The City of Light on the Plaza of Light at the New York World's Fair, 1940/ presented by the Consolidated Edison System Companies. New York: The Company: Distributed by New York World's Fair 1939 Inc. c.1940

Croft, Terrell. Electrical Machinery Principles, Operation and Management. New York: McGraw Hill Book Co. 4th Ed. 1938

Department of Commerce, Bureau of the Census, Wm. J. Harris, Director. Central Electric Light and Power Stations and Steam and Electric Railways with Summary of the Electrical Industries 1912. Washington: Government Printing Office, 1912

Dalzell, Frederick. Engineering Invention, Frank J. Sprague and the U.S. Electrical Industry. Cambridge, MA. The MIT Press, 2010

Freeman, Charles Yoe. The Miracle of Electric Light and Power. New York: Newcomen Society in North America, 1952.

Gilmartin, Gregory F. Shaping The City, New York and the Municipal Art Society. New York: Clarkson Potter, 1995

Graham, Frank D. Audel's New Electric Library, Ten Volumes, for Engineers, Electricians, All Electrical Workers, Mechanics and Students. New York, N.Y. and Indianapolis: Ind. Theodore Audel & Co. Div. of Howard Sams, 8th Ed. 1965

Hawkins and Staff. Hawkins Electrical Guide, Ten Volumes, A Progressive Course of Study for Engineers, Electricians, and Students. New York: Theodore Audel & Co. 1914

International Correspondence Schools. Storage Batteries, Incandescent Lighting, Arc Lighting, Interior Wiring, Modern Electric Lighting Devices, Electric Signs, Electric Heating. Scranton Pa, 1908.

Joint General Committee of National Electric Light Association and Bell Telephone System. Reports on Physical Relations Between Electrical Systems and Signal Systems. New York, 1922

Jones, Payson. A Power History of the Consolidated Edison System. (Compiled as a Reference Work From Original Documentary and Other Sources, with Especial Reference to the Menlo Park and Pearl Street Origins of the System.)New York: Consolidated Edison Co, 1940.

Kimbark, Edward Wilson. Electrical Transmission of Power and Signals. New York: J. Wiley, 1949.

Lamme, Benjamin Garver. Benjamin Garver Lamme, Electrical Engineer, An Autobiography. New York, London: G.P. Putnam's Sons, 1926.

Lieb, John William. The Edison System of Electric Lighting, Presidential Address by John W. Lieb. New York: The Marchbanks Press, 1920.

_____. The Historic Pearl Street, New York, Edison Station. By the First Electrician of the Station Dr. John W. Lieb. New York: 1927

_____. Progress of the Central Station Industry in America, by Dr. John W. Lieb. New York: New York Edison Co., 1928.

BIBLIOGRAPHY

_____. Review of Power Resources and Their Development in the Northeastern United States, Including Statistics and Information Relating to the New England and Middle Atlantic States, Delaware, Maryland, West Virginia, District of Columbia and Virginia and Interconnections with Eastern Ohio. Dr. John W. Lieb. The World Power Conference, London, England, June 30-July 12, 1924. New York: The Marchbanks Press, c1924.

Lincoln, Edwin Stoddard. Electrical Protective Equipment and Power Factor Correction. New York: Essential Books, 1945.

Luce, Charles F. One Hundred Fifty Five Years of Technological Excellence, an address before the Newcomen Society of New York, October 26, 1978. New York: Newcomen Soc., c1979

Martin, Thomas Commerford. The Electric Motor and its Applications by Thomas Commerford Martin and Joseph Wetzler. New York: W.J. Johnston, 1887 (c1886)

_____, 1888.

_____, 1892.

_____, 1895.

_____. Forty Years of Edison Service, 1882-1922; Outlining the Growth and Development of the Edison System in New York City. New York: New York Edison Company, 1922.

_____. The Story of Electricity. Ed. by T.C. Martin and Stephen Leidy Coles. New York: M.M. Marcy, 1924.

Middleton, Robt. G., rev. by Meyers, L. Donald and Tedesco, Joseph A. Audel's Practical Electricity 4th Ed. New York: Macmillan, 1988

Middleton, William D. and Middleton, William D. III. Frank Sprague Electrical Engineer and Inventor. Bloomington, IN: Indiana University Press, 2009

Miller, John A. Modern Jupiter: The Story of Charles Proteus Steinmetz. New York: American Society of Mechanical Engineers, 1958.

Murphy, Herbert A. A Study of Power Factor Costs and Methods of Correction. New York: National Electric Light Association, c1931.

Murray, Thomas E. Applied Engineering. New York: The Ferris Printing Company, 1928.

_____. Electric Power Plants; A Description of A Number of Power Stations... New York, 1910.

_____. Power Stations, by Thomas E. Murray. New York, 1922

Murray, Thomas E. (Jr.) Nuclear Policy for War and Peace. Cleveland: World Publishing Company, 1960

New York Edison Company. Approved Service Equipment by the New York Edison Company, Brooklyn Edison Company...(and others) Effective October 1, 1935. Revisions of Approved Service Equipment Booklet Dated June 1, 1933. New York, 1935.

_____. Big Buildings Using Service of the New York Edison Company. New York: The New York Edison Company, 1926.

_____. Commemorating Fifty Years of Service and the Opening on September 4, 1882 of the Pearl Street Generating Station by Thomas Alva Edison. New York: New York Edison Company, 1932.

_____. The East River Generating Station of the New York Edison Company. New York, 1927.

_____. Electric Automobile Charging Stations and Road Map of New York and Suburban Territory. New York: New York Edison Company, 1912.

_____. 1920.

_____. 1922.

_____. Fresh Air for You. New York: The New York Edison Co., The United Electric Light and Power Co., 1935.

_____. New York in MCMXXIII; An Illustrated Book of the City, Compiled in Honor of the 46th Convention of the National Electric Light Association. New York: New York Edison Company, 1923.

_____. Rates, Service Conditions and Other General Information. March 1909. New York, 1909.

_____. Report of the Electrolysis Surveys of Cable Systems. New York, 1935.

_____. Report of the Electrolysis Surveys of Gas Piping Systems of the Consolidated Gas Company of New York and Affiliated Companies. New York, 1935.

_____. Service and Meter Rules and Regulations, Boroughs of Manhattan and the Bronx. New York: New York Edison Company, 1924.

_____. 1926.

_____. Thirty Years of New York, 1882-1912; Being A History of Electrical Development in Manhattan and

BIBLIOGRAPHY

the Bronx. New York: Press of the New York Edison Company, c1913.

New York State Public Service Commission. In the Matter of Complaints Under Sections 71 and 73 of the Public Service Commissions Law, of Petitioners, Represented by E.F. Jeffe, Inc., against Brooklyn Edison Company, Inc, the New York Edison Company, the United Electric Light and Power Company... (and Others.) Petitioners' Preliminary Presentation as to Facts Concerning the Coal Surcharge. E.F. Jeffe, Inc., Representative for the Petitioners... New York, 1930.

_____. In the Matter of the Hearing On Motion of the Commission as to Rates and Rate Structures of Various Corporations Supplying Electricity in the City of New York and Suburban Territory, Brief Submitted by Silvian Palmer... Representing William Randolph Hearst, United Cigar Stores Co. of America and Other Large Customers of the New York Edison Co. and the United Electric Light and Power Co. New York, 1931.

New York (State) Public Service Commission 1st District. Preliminary Abstracts for Reports of Half Year... of the New York Edison Co., United Electric Light and Power Co. New York, 1907.

O'Neill, John J. Prodigal Genius; The Life of Nikola Tesla, New York, N.Y.: I. Washburn, Inc. 1944,

Polmar, Norman, and Allen; Thomas B. Rickover. New York: Simon and Schuster 1982

Prout, Henry Goslee. A Life of George Westinghouse. Arno Press 1972 (c1921.)

Public Service Commission Case 1533 SM 5563-74 Hylan vs. New York Edison Company, Exhibit and Testimony of Matthew S. Sloan, Description of Changeover of Direct Current Electrical Service to Alternating Current, April 30, 1929

Seifer, Marc. Wizard, the Life and Times of Nikola Tesla: Biography of a Genius. Seacaucus N.J.: Carol Pub. c1996.

Sloan, Matthew Scott. Consolidations in the Electric Utility Industry. New York: National Electric Light Association, c1929.

Smith, Frank W. The Case for Private Enterprise; An Address Delivered by Invitation at Princeton University on December 12, 1935, Before A Joint Meeting of the American Whig Society and the Cliosophic Society of Princeton by Frank W. Smith. Princeton, N.J.: Princeton University Press, 1936.

Sprague, Frank J. Electric Motors. Annapolis, Maryland, 1887.

_____. Some Applications of Electric Transmission. A Lecture Delivered Before the Students of Sibley College. New York: Sprague Electric Railway and Motor Co., 1887.

_____. Transmission of Power by Electricity. Discussion Before the National Electric Light Association, Philadelphia, Pa. February 16, 1887. New York: Sprague Electric Railway and Motor Co., 1887.

Sprague, Harriet. Frank J. Sprague and the Edison Myth. New York: The William-Frederick Press, 1947.

Starr, Tama and Hayman, Edward. Signs and Wonders; The Spectacular Marketing of America. New York: Currency-Doubleday, 1998

Stratemeyer, Edward. Bound to be an Electrician, or Franklin Bell's Success. New York: William L. Allison & Co. c1897.

Steinmetz, Charles Proteus. General Lectures on Electrical Engineering by Charles Proteus Steinmetz. Comp. and Ed. by Joseph Leroy Hayden. New York: McGraw Hill, 1918.

Steinmetz, Charles Proteus. Theory and Calculation of Alternating Current Phenomena, by C.P. Steinmetz with the Assistance of E.J. Berg. New York: W.J. Johnston Co., 1897.

Stubbing, George Wilfred. Power Factor Problems in Electricity Supply; an Elementary Treatise for Students, Supply Engineers, and Power Users. London: E, & F.N. Spon Ltd.; New York: The Chemical Publishing Co. Inc. 1940.

Tesla, Nikola. Transmission of Power: Polyphase System, Tesla System. Pittsburgh, Pa.: Westinghouse Electric and Manufacturing Co., 1893.

Waddicor, Harold. The Principles of Electric Power Transmission by Alternating Current. London: Chapman & Hall Ltd. 1928.

Wainwright, Nicholas B. History of the Philadelphia Electric Company, 1881-1961. Philadelphia, 1961.

Walters, Helen B. Nikola Tesla, Giant of Electricity. New York: Crowell, 1961.

Westchester Lighting Company. The Story of Industry in Westchester County; A Unique Radio Series. Mount Vernon, N.Y.: Westchester Lighting Co., the Yonkers Electric Light & Power Co., 1948.

Westinghouse Electric Corporation. Electrical Transmission and Distribution Reference Book by

Central Station Engineers of the Westinghouse Electric and Manufacturing Co. East Pittsburgh, Pa.: The Company, 1944.

Wilson, Charles Edward. More Power for America. General Electric, 1950.

Woodbury, David Oakes. Beloved Scientist; Elihu Thomson, a Guiding Spirit of the Electrical Age, by David O. Woodbury with a foreword by Owen D. Young. New York, London: Whittlesey House, McGraw-Hill Book Company, Inc. 1944.

PERIODICALS:

Around the System (Published by the Consolidated Edison Company 1936-1987)

Bulletin of the New York Edison Company.

Electric Journal (Published by the Westinghouse Electric and Manufacturing Company, 1904-1939)

Electrical World (McGraw Hill, University of Michigan Microfilm)

Electrical World Directory of Utilities. (McGraw Hill)

Edison Monthly (Published by the New York Edison Company)

Edison Weekly (Published by the New York Edison Company)

General Electric Review

United Service (Published by the United Electric Light & Power Co.)

ANNUAL REPORTS:

Consolidated Edison Company

Consolidated Gas Co. Units; New York Edison Co. and United Electric Light & Power Co.

INDEX

Page references in *italics* indicate photographs.

A

Abercrombie & Fitch, 130
advertising: for electricity, 133–135, 173–174; lighted signs, 28, 72, 79, 113–114, 130–131, 134–135; neon signs, 172; postwar, 172
Affiliated Companies, 64–65
air conditioning, 176
Allis-Chalmers, 126
alternating current, 4; changeover to, 147–152; comprehensive systems, 44–47; customer system considerations, 49–52; distribution, 49–52, 135–136, 160–161; generation, xiv, 37–59; incursion, 107–143; lagging current, 50; leading current, 51; monocyclic systems, 46–47; national trend, 123–126; power generation, 114–116; railway systems, 56–59; "The Theory and Operation of Alternating Current" (Cunningham), xv; transformer (sub) stations, 116–122; transmission, 37–59; triumph of, 145–157

alternating current motors, 42–43, 126
Ann Street (New York City, NY), 14
apparent power, 50
arc lamps, 5, 90
arc light systems, 3–4
Ashton Iron Works, 11–12
Atomic Energy Commission, 181, 184
"At Your Service" slogan, 78
automatic distribution networks, 137–138, *139,* 139–140
Automat restaurants, 161
automobiling, 171

B

B. Altman's, 72
Baltimore & Ohio Railroad, 2
basement substations, 72
battery support, 31–32
Bay Ridge Station (Brooklyn, NY), 163
Beekman Street (New York City, NY), 14
Bell, Franklin, 61
Bellevue Hospital, 114
Bergmann, Sigmund, 12
Bergmann Company, 12
Blackstart capability, 189
Blizzard of '88, 21
Boston Edison Company, 146
"Bound to Be an Electrician, or, Franklin Bell's success" (Stratemeyer), 61

INDEX

Boy Scouts of America, 83
Bradley, Charles, 45–46
Bradley Electric Company, 45–46
Brady, Anthony N., 62–66
Broadway (New York City, NY), 1–2, 9, 28, 89
Broadway Cable, 55–56, 67
Bronx Civic Center, 80–81
Bronx District, 87
Bronx station (Port Morris), 97
Brooklyn, New York, 53–54, 148–149
Brooklyn Bridge, 1, 87
Brooklyn Edison, 95–96, 142, 149, 153, 162–164, 170; Hudson Avenue Station, 128, 143; interconnection with United and New York Edison, 127–129; merger with NY & Queens Co., 159–160
Brooklyn Rapid Transit Company, 63, 89, 153
Brown-Boveri Company, 178
Brunswick Hotel, 10
Brush, Charles, 3, 8–10
Brush Electric Company, xvii, 3, 9, 15, 107
Bryant Park, 171–172
building networks, 176

C

California Electric Light Company, 3
Canal Street (New York City, NY), 89
Cataract Commission, 48, 190
Cedar Street (New York City, NY), 14
central stations, 9–35

Central Telegraph & Electrical Subway Company, 63, 110
Centre Street (New York City, NY), 89
chemical meters, 15
Chicago, Illinois: basement substations, 72; Columbian Exposition (1893), 48, 107–108, 190; Field Building, 141
Christmas tree lights, 7–8, 62
Chrysler Building (New York City, NY), 141
circuit breakers, 98–99
Citizens Electric Illuminating Company, 51
Clark, Edith, 129
coal, 92, 188
Cold War, 173–174
College of St. Ignatius, 3
Columbian Exposition (Chicago, 1893), 48, 107–108, 190
Columbia Presbyterian Medical Center, 140
Commonwealth Edison Company, 136, 146–149, 183
commutators, 4
"A Compact Alternating Current City Substation," 130–131
condensers, synchronous, 51
Consolidated Edison Company Inc., 159–160, 163, 168, 172–174; advertising, 173–174; expansion, 174; "Lights On" slogan, 172; nuclear generating station, 183–184; peak loads, 176; power stations, 180
Consolidated Gas Company, 63–64, 159–160, 170–171
consolidation, 159–168
Cooper, F.G., 78

cost benefits, 7, 173. *see also* rates for service
Crompton-Howell, 32
Cumberland Hotel (New York City, NY), 28
Cunningham, Joseph (Joe), xv
customer satisfaction, 93–94, 146
customer service: "At Your Service" slogan, 78; changeover, 160–164; demand for power, 83–85, 95–96; direct current, 91–105; peak demand, 31–32, 176

D

Daily News, 176
Davenport, Thomas, 2
DaVinci, Leonardo, 78
DeLaval turbine, 34
Delmonico's Restaurant, 71
demand for power, 83–85, 95–96; peak demand, 31–32, 176
department stores, 72
"Dig We Must For A Growing New York," 175
direct current, 4, 186; changeover from, 147–152; customer service peak, 91–105; demise of, 145–157, 187–188; disconnection of last customers, 187–188; distribution capacity, 160–161; history of, 190; Pearl Street station, xiv; persistence, 186–187; selection of, 52–54; substations, 54–56; system limitations, 33–35
distribution: automatic networks, 135–137, *139,* 139–140; improvements in, 97–99; New York network, 137–138; overhead systems, 136–137; underground lines, 136–137; United network, 137–138

distribution stations, 52. *see also specific stations*
distribution transformers, 125–126
domestic appliances, 92–93, 133–134, 173
double conversion, 55
Drexel, Morgan & Company, 13–14
dual contracts, 132
Duane Street Station (New York City, NY), 33–34, 55–56, 89, 96, 166
Duncan, Louis, 42
Duprez, Marcel, 190
Duquesne Power and Light Company, 183
dynamos, 4–5
Dynamotor converters, 178

E

East 12th Street Station (New York City, NY), 33–34, *34*, 55–56, 186
East 26th Street Station (New York City, NY), 73, 186
East 29th Street Station (New York City, NY), 107–109, 114–115
East 32nd Street Station (New York City, NY), 99–100
East 39th Street Station (New York City, NY), 71, 100, 150
East 41st Street Station (New York City, NY), 100
East 96th Street Station (New York City, NY), 56–57
East River station (New York City, NY), 96
Edison, Thomas, xiv, xvii–xviii, 5–8, 10–15, 21–22, 25, 63, 187–188; concerns for safety, 45; 3-wire system, 17
Edison Bulletin, 78

INDEX

Edison Company of Brooklyn, 87
Edison Electric Illuminating Company, 10–11, 15, 21–25, 33–34, 52–53, 63–65, 94, 110–111, 153; electric kitchen, 28–29; Emergency Department, 72; expansion, 29–31; midtown power stations, 25–27, 26; motor inspection bureau, 20; Second District, 72–73; total capacity, 29–30. *see also specific stations*
Edison General Electric Company, 37, 45
Edison Machine Works, 11–12
Edison Monthly, 78–79
Edison Shafting Company, 12
Edison Tube Company, 12
Edison United Manufacturing Companies, 21, 37
Edison Weekly, 78, 105
Ediswan, 6
Eickemeyer Company, 46
Eisenhower, Dwight, 182
"El Bari Gin Rickey" sign, 113–114
Electrical Age, 28
electrical service: consumer usage of, 93; direct current, 91–105; growth through economy of scale, 61–90; interruptions, 30; New York network, 137–138; public services, 8; rates for, 30; single-phase, 110; spikes (peaks), 31–32; system expansion, 95–97; system limits, 16–18; two-phase, 110; variation in, 102–103
electrical steel, 61–62
Electrical World, 8, 84–85, 121, 129–130, 140, 145, 155, 178
electric appliances, 92–93, 133–134, 173
electric cables, 175

electric irons, 112
The Electric Journal (Westinghouse Company), 39, 136–137
electric kitchens, 28–29, 85, 93
electric lighting, 62, 80–81; arc lamps, 5, 90; arc light systems, 3–4; Christmas tree lights, 7–8, 62; fluorescent lamps, 172; general electric metalized ("gem") lamps, 62; "Hylo" or "turn down" lamps, 82; incandescent lamps, 90; indoor electric light, 10–16; neon signs, 172; single-phase service, 110; street lights, 90; transformation of urban life by, 28–29
electric power: apparent, 50; beyond lighting, 18–20; costs of, 7; demand for, 31–32, 83–85, 95–96, 176; generation of, 114–116; indoor lighting, 10–16; private, 1–8; private plants, 93–94; reactive, 50; transformation of urban life by, 28–29
electric power stations, 180; alternating current transformer (sub) stations, 116–122; basement substations, 72; capacity improvements, 98–99; central stations, 9–35; direct current substations, 54–56; distribution stations, 52; first, 9; receiving stations, 24; transformer stations, 135; vertical style stations, 33–34. *see also* substations; *specific stations*
Electric Shops, 133–134, *134*
Electric Show, 28–29
electric signs, 28, 72, 79, 113–114, 130–131, 134–135; planning kits for, 92–93
Electric Vehicle Handbook (Smith), 171
Electric World, 107–108

elevators, 1, 32–33, 94
Elizabeth Street station (New York City, NY), 109, 114–117, 131
employee protection plans, 170–171
environmental improvements, 89–90
Excelsior Company, 177–178

F

FACTS (Flexible Alternating Current Transmission Systems), 191–192
Fairbanks-Morse, 126
Fall River, Massachusetts, 146
Farmer, Moses G., 17
Father Knickerbocker, 78
Federal Archives (Washington, DC), 141
Federal Power Commission, 167–168
Ferry Street (New York City, NY), 11
Field Building (Chicago, IL), 141
Fifth Avenue (New York City, NY), 7, 54–55
Fisk Tire, 130–131
Flexible Alternating Current Transmission Systems (FACTS), 191–192
fluorescent lamps, 172
Ford, Henry, 166
Fortescu, Charles, 129
Foster & Biel's Music Hall, 10
Franklin Institute, 4–5
Fulton, Robert, 79
Fulton Street (New York City, NY), 12–14

G

Gamewell alarm systems, 66–67
gas, 188
Gas Statutes, 11
Gaulard, Lucien, 39
"gem" (general electric metalized) lamps, 62
General Electric Company, 45–49, 96, 113; *GE Review*, 117; lamp factory, 82; postwar, 173–174; research efforts, 126
general electric metalized ("gem") lamps, 62
GE Review, 117, 119–120
giants, 153–157
Gibbs, John, 39
Gilsey House, 10
Gimbel's department store, 71, 100–101
Gimbel's Substation, 71–72, 100–101
Glenwood, 97
Goerick Street (New York City, NY), 12
Gold Street station (Brooklyn, NY), 163, 179–180
Gold Street station (New York City, NY), 177–178
Gorham Company, 100
Grand Central Palace, 71
Great Barrington, Massachusetts, 24, 38–39
Greenfield Village (Detroit, MI), 166

H

Hamilton Place (New York City, NY), 112
Hardy Boys series (Stratemeyer Syndicate), 61
Heatherbloom Petticoat "Silk's Only Rival" sign, 113–114

heat rate, 116
Hell Gate station (New York City, NY), 95–97, 127–128, 153, 162, 185
Hippodrome Arena, 29
Holborn Viaduct station (London, England), 14–15
Hopkinson, John, 16–17
Horn & Hardart, 161
household appliances, 92–93, 133–134, 173
Houston, Edwin, 4
Hudson, Hendrick, 79
Hudson Avenue Station (New York City, NY), 128, 143
Hudson & Manhattan (H&M) Railroad, 59, 95
hydraulic elevators, 1
"Hylo" or "turn down" lamps, 82

I
incandescent lamps, 90
Indian Point nuclear station, 183–185, 189
indoor electric light, 10–16
induction coil, 37–38
induction motors, 40–41, 126
Insull, Samuel, 55
Interborough Rapid Transit Company (IRT), 181

J
Johns Hopkins University, 42
Johnson, Edward H., 7–8, 17, 19
John Street (New York City, NY), 14

K
Kay-Scherer Surgical Instrument Company, 113
Kensico Dam, 87

L
Ladies' Mile, 1–2, 9, 74, 102–103
lagging current, 50
Lamb and Poole, 81
Lamme, Benjamin G., 41–43, 46, 51, 108
Lamme, Bertha, 41
lamps, 14–15, 81–82; arc lamps, 5, 90; Christmas tree lights, 7–8, 62; fluorescent lamps, 172; General Electric Company factory, 82; general electric metalized ("gem"), 62; "Hylo" lamps, 82; incandescent, 90; "turn down" lamps, 82; Westinghouse Lamp Company, 131–132
leading current, 51
Levey, 113–114
Lieb, John W., 13, 77–78, 97–98, 105, 125, 142, 156–157
light bulbs, 81–82
lighted signs, 28, 72, 79, 113–114, 130–131, 134–135
lighting: arc lamps, 5, 90; arc systems, 3–4; Christmas tree lights, 7–8, 62; fluorescent lamps, 172; general electric metalized ("gem") lamps, 62; "Hylo" or "turn down" lamps, 82; incandescent lamps, 90; indoor electric, 10–16; neon signs, 172; single-phase service, 110; soft light, 5–8; street lights, 90
lighting plants, 5
"Lights On" slogan, 172

Lights & Shadows, 111
LIRR. *see* Long Island Rail Road
"Live Better Electrically" slogan, 173
load factor, 18
Loew, Marcus, 131–132
Loew's Theater, 131–132
London Underground, 19
long-distance transmission, 189–190
Long Island City, New York, 129
Long Island Rail Road (LIRR), 59, 163–164, 169
Lord & Taylor, 165
Los Angeles, CA, 145–146
Louisville, KY, 145

M
Macy's, 72
Madison Park, 1–2, 9–10
Manhattan Beach Hotel (Brooklyn, NY), 28
Manhattan Electric Light Company, 34, 54–55
manhole switches, 136
Maria (of Rumania), 96
McCoy, W.E., 107–108
mercury arc rectifiers, 177–178
metal filament "gem" (general electric metalized) lamps, 62
meter men, 31
Metropolitan Device Co., 154
Metropolitan Life Building, 73, 80–81
Metropolitan Opera House, 29

Metropolitan Railway Company, 56–58
Metropolitan Street Railway, 56–57, 87
milestone year (1962), 177–180
modernization, 161, 185
monocyclic systems, 46–47, 126
Morgan, J.P., 45
Morgan Crucible Company, 131–132
Municipal Electric Light Company, 37–38
Murray, Thomas E., 63–66, 85–86, 90, 141–142, 153–156
Murray, Thomas E., Jr., 181–182
Murray Hill Building, 101
Murray Radiator Co., 154–155

N
Nassau Street (New York City, NY), 11
National Biscuit Company, 109, 113, 121
natural gas, 188
NBC, 166
neon signs, 172
Neri, Joseph, 3
Newark, New Jersey, 145
New Deal, 167–168
Newport Naval Base, 23
newspaper advertising, 134–135
New York, New Haven & Hartford Railroad, 162
New York Central Railroad, 59, 97, 169
New York City, New York, xvii–xix; alternating current transformer (sub) stations, 116–122; basement

substations, 72; central stations, 9–35; changeover to alternating current, 147–152; direct current substations, 54–56; distribution capacity, 160–161; first electric power station, 9; indoor electric light, 10–16; midtown, 25–27, 130–133, 164–165; private power, 1–8; regional interconnections, 127–129; substations, 71–73, 88; system expansion, 95–97; system limits, 16–18; World's Fair (1939), 166. *see also specific stations*

New York Edison Co. Inc., 159–160

New York Edison Company, 63–64, 81, 84, 105, 130, 142, 146, 170; Bronx District, 87, 129; changeover to alternating current, 149–151, 162; customer service, 143, 162; direct-current distribution, 160; marketing, 78–81; merger with United, 159–160; morale, 82–83; number of structures, 132–133; publications, 78–79; railway power supply, 88–89; rate reduction plan, 91; regional interconnection with United and Brooklyn Edison, 127–129; sales campaign, 92–93; slogan, 78; substations, 81, 103–105, 132–133. *see also specific stations*

New York Electric Equipment Company Limited, 29–30

New York Gas and Electric Light Heat & Power Company, 65

New York Herald, 14

New York network, 137–138

New York Public Library, 80–81

New York & Queens Electric Light & Power Company, 95–96, 129, 159–160, 170

New York State Public Service Commission, 152, 187
New York Times, 72, 79, 171
Niagara Falls, New York project, 47–49, 190
night life, 28
nuclear age, 180–184
nuclear generating stations, 182–184, 189

O
oil, 188
Oil Embargo (1973), 189
Oil Embargo (1979), 189
Omaha, Nebraska, 145
133 Hz systems, 107
140th Street-Rider Avenue station, 87
O'Neill, John J., 190–191
overload protection, 98–99

P
Pacific Building (San Francisco, CA), 3
Panic of 1907, 122
Pantaleoni, Diomede, 39
Pantaleoni, Guido, 38–39
Paramount Building, 72
Park Theater, 10
PATH System, 95
peak demand, 31–32, 176
Pearl Street (New York City, NY), xiv, xvii, 12, 14–15
Pearl Street Station (New York City, NY), 30, 33–34, 166, 187–188

Peck Slip (New York City, NY), 11
Pen Corner, 132
Penn Street Station (New York City, NY), 37–38
Pennsylvania Railroad, 59
Peoria, Illinois, 136
Philadelphia, Pennsylvania, 147–148
Philadelphia Centennial Fair (1876), 4–5
Philadelphia Electrical Exposition (September, 1884), 19
Philadelphia Electric Company, 146, 148–149, 183
Pine Street (New York City, NY), 14
pioneers, 1–8
Portland, Oregon, 48, 145–146
Port Morris, 97
postwar development, 172–174
Powell, Charles, 171–172
power factor, 50
power plants, 188–189; private plants, 93–94; public services, 8–9. *see also* electric power stations
power sources, 188
prices: Breakdown rate, 30; coal surcharges, 92; cost benefits, 7; Full Central rate, 30; rates for service, 30, 91
private plants, 93–94
private power, 1–8
promotion and advertising, 133–135
public services, 8–9
Public Utility Holding Companies Act of 1935, 167–168

Q
Queens, New York, 174

R

railways, 56–59, 88–89; street railways, 32–33. *see also specific railways*
rates for service, 30, 91
reactance, 50
reactive power, 50
receiving stations, 24
recent change and developments, 185–193
rectifiers, 177–180, 186
redevelopment, 2
Redlands, California, 47
refrigeration, 140–141
regional interconnections, 127–129
Rickover, Hyman, 180–183
riser cables, 141
Riverside Drive Viaduct, 80–81
"Roman Chariot Race" sign, 72
Roosevelt, Franklin, 167–168
Root, F.S., 146
rotary converters, 41–44
Ryan "Scintillator" Searchlights, 80

S

safety, 45, 154
Sak's, 72
San Diego, California, 145
San Francisco, California, 3, 145–146
Scott, Charles, 48–49
scrubbers, 89–90

INDEX

Seattle, Washington, 146
72nd Street Station (New York City, NY), 54–55
Shallenberger, Oliver, 43–44
Sherman Creek Station (New York City, NY), 122–123, 135, 162, 171, 185
Sherry's Restaurant, 71
Siemens Company, 45
Siemens-Halske, 45
sign planning kits, 92–93
signs: lighted, 28, 72, 79, 113–114, 130–131, 134–135; neon, 172
silicon rectifiers, 179–180
"Silk's Only Rival" sign, 113–114
Sixth Avenue (New York City, NY), 1–2
60 Hz systems, 107–108, 110–111, 124–125, 127–129
S-K-C (Stanley-Kelly-Chesney) Company, 45, 67
skyscrapers, 32–33
Sloan, Matthew, 142–143, 149–151
Smith, Alfred E., 171–172
Smith, Frank W., 141–142, 156, 170–172
social intervention, 167–168
soft light, 5–8
Soldiers and Sailors monument, 80
spikes (peaks) in service, 31–32
spinning, 124
Sprague, Frank Julian, 17–20, 22–25, 32–33, 37, 58
Sprague motor firm, 37
Spruce Street (New York City, NY), 11, 14

St. Louis World's Fair (1904), 166
St. Patrick's Cathedral, 80–81
Stanley, William, 24, 38–39, 45
Stanley-Kelly-Chesney (S-K-C) Company, 45, 67
Staten Island, 111, 174
Staten Island Edison Company, 159–160
stations: central stations, 9–35; distribution stations, 52; first, 9; receiving stations, 24; transformer stations, 135; vertical style, 33–34. *see also* substations; *specific stations*
steam engines, 7
Steinmetz, Charles Proteus, 46, 126
Steinway emporium, 10
Stevens Institute of Technology, 13
Stilwell, Lewis B., 58
Stratemeyer, Edward, 61
Stratemeyer Syndicate, 61
street lights, 90
street railways, 32–33
Structural Steel Board, 152, 187
Stuyvesant Hotel, 10
substation operators, 92
substations, 49, 52, 67, 71–73, 88–89, 99–101; alternating current transformer (sub) stations, 116–122; basement, 72; capacity improvements, 98–99; direct current, 54–56; elimination of, 135; operation, 74–75, 92, 103–105; railway system, 58–59. *see also specific stations*
subway power, 89

Swan, Joseph, 6
Sweet's restaurant, 14
synchronous condensers, 51
system operators, 66–67, 104

T
"Talking Machine" sign, 113–114
Teaser circuit, 126
television, 173
Telluride, Colorado, 48
Tenderloin, 2
Tesla, Nikola, 7, 39–44, 48, 190–191
"The Theory and Operation of Alternating Current" (Cunningham), xv
Third Avenue Railway, 88–89
Thomson, Elihu, 3–5, 8, 37–38, 42
Thomson-Houston Company, 4, 10, 45–46
3-wire systems, 17
Thury, Rene, 190
Time, 113–114
Times Square (New York City, NY), 71–72, 111, 165
Tin Pan Alley, 2
tradition, 105
transformers, 38–39
transformer stations, 135
transmission improvements, 97–99
Truman, Harry S., 181
"turn down" lamps, 82
25 cycle flicker, 43–44

U

"U-Need-A Biscuit" sign, 113
Union League Club, 27
Union Square (New York City, NY), 33–34
Union station, 110–111
United Electric Light & Power Company, 39–40, 52–53, 64, 81, 87, 107–114; alternating-current distribution, 160; automatic distribution network, 135–138, *139,* 139–140; changeover to alternating current, 149–150; corporate office, 121–122; customer accounts, 121; customers, 132; Electric Shops, 133–134, *134*; electric signs, 113–114; expansion, 128–133; incursion, 113–114; innovation, 141–142; merger with New York Edison, 159–160; number of buildings, 132–133; 133 Hz system, 107; promotion and advertising, 133–135; rate reduction plan, 91; regional interconnection with New York Edison and Brooklyn Edison, 127–129; 60 Hz systems, 107; under Smith, 170–171; System Operator, 121–122. *see also specific stations*
United network, 137–138
United Service, 133–134
United States Illuminating Company, 39–40, 141–142
United States Naval Observatory, 135
United States Patent Office, 46
University of San Francisco., 3
urban life, 28–29
U.S. Atomic Energy Commission, 181, 184
U.S. Commerce Department, 124

U.S. Navy, 79–80, 183
U.S.S. Nautilus, 180–181

V
Vertical style stations, 33–34
Victor "Talking Machine" sign, 113–114
voltage drops, 16
voltage regulation, 98–99

W
W2XBS, 166
Wabash, Indiana, 3
Wagner and Reliance, 126
Wallabout Market (Brooklyn, NY), 149
Wall Street (New York City, NY), 11
Wanamaker's Department Store, 84–85
Warren Self Winding Clock Company, 135
Washington, D.C., 146
Washington Square, 80–81
Waterman Building, 132
Waterman Pens, 132
Waterside complex, 65–67, 95–97, 114, 121–122, 189; alternators, 117; expansion, 174–175; reduction, 185
Waterside Generating Station Number One, 65–67, 68–69, 69–71, 86–88, 115–116
Waterside Generating Station Number Two, 68, 85–88, 96, 114–115, 162–163
weather issues, 21–22

West 16th Street Station (New York City, NY), 73–74, 76–77, 186
West 17th Street Station (New York City, NY), 186
West 22nd Street Station (New York City, NY), 160, 186
West 24th Street Station (New York City, NY), 131
West 26th Street Station (New York City, NY), 25–32, *26,* 72–74, 92, 101–102, 113, 164–165, 185–186
West 27th Street Station (New York City, NY), 73
West 35th Street Station (New York City, NY), 186
West 36th Street Station (New York City, NY), 165, 186
West 39th Street Station (New York City, NY), 25, 27, 29–30, 34, 55–56, 71–72, 101, 185–186; changeover to alternating current, 150; substations, 71–73, 99–101
West 41st Street Station (New York City, NY), 72
West 45th Street Station (New York City, NY), 130–131
West 47th Street Station (New York City, NY), 186
West 48th Street Station (New York City, NY), 100, 186
West 53rd Street Station (New York City, NY), 32
West 97th Street Station (New York City, NY), 131, 137
West 146th Street Station (New York City, NY), 114, 116–118, *118,* 119–120
Westchester, New York, 159–160
Westchester station (Glenwood), 97
Westinghouse, George, xiv, 38–40, 47–48, 170, 182
Westinghouse Current, 41

INDEX

Westinghouse Electric and Manufacturing Company, 39–51, 113, 137–138; Columbian Exposition exhibit (Chicago, 1893), 190; comprehensive alternating current system, 44–45; distribution transformer, 125–126; *The Electric Journal,* 39, 136–137; Niagara project, 48–49; railway systems, 58–59
Westinghouse Lamp Company, 131–132
Williamsburg, Brooklyn, 37–38, 108
wiring, 147
women substation operators, 92
Wood, James J., 4–5
World's Fair (New York City, 1939), 166
World's Fair (St. Louis, 1904), 166
World War I, 145
World War II, 92, 168; postwar prosperity, 169–184
Worth Street (New York City, NY), 14
Wrigley's Gum, 130

Y
Yonkers, New York, 159–160

Made in the USA
Middletown, DE
17 February 2024